German Aviation Industry in WWI

Volume 1

Michael Düsing

German Aviation Industry in WWI

Volume 1

Michael Düsing

Great War Aviation Centennial Series #84

Acknowledgements

Color aircraft profiles © Bob Pearson. While every care is taken, some colors are speculative based on known practices at the time. Purchase Bob's CD of WWI aircraft profiles for $50 US/Canadian, 40 €, or £30, airmail postage included, via Paypal to Bob at: bpearson@kaien.net

For information on our aviation books, please see our website at: www.aeronautbooks.com. Aeronaut Books is looking for photographs of rare German aircraft of WWI for our books. To help please contact the publisher at jherris@me.com.

Interested in WWI aviation? Join The League of WWI Aviation Historians (www.overthefront.com), Cross & Cockade International (www.crossandcockade.com), and Das Propellerblatt (www.propellerblatt.de).

ISBN: 978-1-964637-00-6

© 2024 Aeronaut Books, all rights reserved
Text © 2024 Michael Düsing
Design and layout: Jack Herris
Cover design: Aaron Weaver
Digital photo editing: Jack Herris

Books for Enthusiasts by Enthusiasts
www.aeronautbooks.com

Table of Contents

Volume 1

1. Acknowledgements	4
2. Foreword	5
3: Development of the Aviation Industry During the First World War	8
4: Development of German Aircraft Factories During the First World War	67

Volume 2

4: Development of German Aircraft Factories During the First World War (continued)	4
5: Development of the Aircraft Engine Industry	102
6: Development of the Propeller Industry	183
7. Bibliography	260

1. Acknowledgements

Any work First, I would like to thank Jack Herris for inspiring me to undertake this project of a book. In addition, Jack deserves my thanks for the large number of pictures that he has supplemented to this book.

As before, I would like to thank the many supporters. I would like to mention Koloman Mayrhofer (Pitten, Austria), Siegfried Seidel and many others.

Finally, I would like to thank my wife Ilona, who for many months only saw me absent-mindedly at my desk.

2. Foreword

Above: Proud crew of a frontline aircraft.

With the emergence of the first powered aircraft, the idea of using this new technology as a means of transportation did not necessarily arise. The first years of the aircraft's existence were reserved for sport, general amusement at air shows and self-affirmation for a few pilots. Nevertheless, the seed was sown by the Wright brothers and spread around the globe at a rapid pace.

But what should be the mission of aircraft? History has shown that all efforts to develop and use aircraft for civilian purposes have been overtaken by the requirements of wartime use.

Even though airmail stamps appeared before the war, proving that the airplane was used for civilian traffic and postal service, in reality this meant little more than the existence of a few sporadic attempts. At first, the airplane did not get beyond occasional use. Despite all the enthusiasm among the proponents of aviation, there was a lack of investors and engineers to take up the challenge. Science was in its infancy and lagged behind practical aviation. So the pilot of an airplane was considered the hero of the hour as the adversary of all adverse circumstances.

Before the 1st World War, the possibility of commercial passenger transport was also not considered. The sporting competitions were more or less aimed at achieving new national and world records. Further, higher, faster, with or without passengers - these were the challenges to the aircraft manufacturers of the first hour, before the military administrations of the respective countries supplied the existing aircraft industry with orders on a large scale.

Nevertheless, even during the World War there was no lack of isolated attempts to use the airplane for peaceful purposes, e.g., in the last period of the war the establishment of a German field mail air service from Reval (former German name, now Tallinn, capital of Estonia) to Odessa and from Vienna to Kiev. But these attempts concerned only organizational beginnings, by no means aeronautical ones, because the operation was accomplished during its short existence only with military airplanes.

The 1st World War, which saw tens of thousands of aircraft gradually enter service on each front, gave technology an unprecedented opportunity to develop aircraft. In the process, of course, the idea of training the aircraft for the

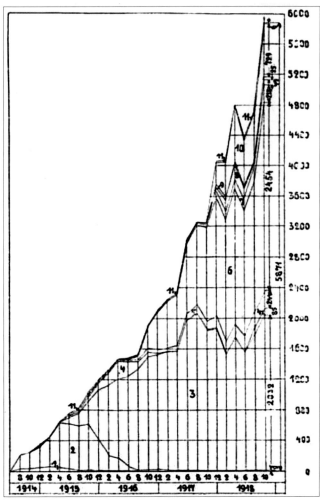

Above: Stock of German front-line aircraft.
Source: IACC report, 1920.

service of peaceful traffic was completely pushed into the background. It was much more important for the war aircraft to bring their combat capability to a maximum.

Technological jumps were felt daily, with pilots achieving new peaks of achievement based on the industry's efforts to increase the performance of flying machines. Initially, no attention was paid to aerodynamics. The wire entanglements of the kite boxes were history by the beginning of the World War. These draggy constructions were replaced by modern and efficient constructions within a very short time.

Aircraft manufacturers were not the only ones to play a major role in this rapid development. The engine industry also made a not insignificant contribution. The most successful companies later in the war were those that were already active in the automotive industry, especially companies such as Daimler, Benz, Adler and others.

The idea of human flight is ancient. The practical solution of this task, namely to lift oneself purposefully from the ground, was investigated early on, but due to the lack of technical aids, these experiments were mostly carried out on paper, or they failed miserably at the first attempt. It was not until the beginning of the 20th century that one was able to solve the problem of human flight, after the entire technology had been developed to such an extent that all the necessary preconditions concerning materials, propulsion systems, meteorological and, above all, basic aerodynamic knowledge were fulfilled.

After the demands made on the Luftwaffe during World War I had brought the development of the airplane, which at the beginning of the war was not much more than a highly fragile and unreliable piece of sporting equipment, to a level foreseen only by a few, a parallel development had to begin after the end of the war, aimed at the usability of the airplane for business, trade and transport, utilizing the experience gained during the war.

Within the period covered by this book, the industry has managed to develop aircraft that can carry 6 to 10 people. It must also be pointed out that, in addition to the familiar materials such as wood, fabric and wire, new materials such as metal were used. The industry developed from the single-seat single-, double- or multi-decker of about 300 kg take-off weight to the large aircraft of several tons. Speeds increased from about 50 km/h to 200 km/h within the first 10 years with the Fokker D.VIIF.; by the end of 1918, people were venturing to altitudes of over 9,000 meters instead of staying at 15 meters.

The first beginnings were made in 1914, when Claude Dornier developed metal aircraft based on a suggestion by Count Zeppelin. From the outset, the materials used were steel and duralumin in mixed construction. Dornier's development subsequently revealed ever more perfect applications. All-metal hulls were developed in 1915, and all-metal fuselages in 1917. While the first aircraft in the founding years usually flew only when there was no wind in the morning or evening, they were able to fly at night just a few years later.

This book deals exclusively with the development of the German aircraft industry from its beginnings to the end of the First World War. Although the development of airships and balloons is deliberately omitted, it must be noted that the immense development of a new branch of industry, such as the aircraft industry, cannot be covered in just one book. The intention is to focus on the development in general and the companies involved.

Great Britain and the German Empire built the largest air fleets within their respective power blocs during World War I, and by the end of the war had the largest aviation industry in terms of monthly aircraft output. On the Allied side, France had a similarly large aeronautical industry, Italy and Russia significantly smaller, and the USA, which had entered the war in April 1917, was only just getting started when the armistice was signed in the fall of 1918. Among the Mittelmächte[1] (Central Powers), Germany provided by far the largest design and manufacturing capacity of airframes and engines; production in Austria-Hungary remained limited for various reasons.

Conceptually, the aeronautical industry includes the

manufacturers of airframes, aircraft engines, propellers and accessories as well as the manufacturers of balloons and airships, but this paper will concentrate on the airframe manufacturers of "heavier than air" aircraft and will only touch on the other areas.

The industrialization of aviation led to a division into specialized companies, and there was a predominant organizational and technical separation between airframe, aircraft engine and accessory production, a threefold division that could already be observed at the General Aircraft Exhibition (ALA) held in Berlin in April 1912.

The development of the aviation industry, military aviation and civil aviation involved a large number of players with different, occasionally strongly divergent views and interests; neither "industry" nor "government" formed a homogeneous block.

---

1 Mittelmächte: The German Reich and the states allied with it during WWI were referred to as the "Central Powers".

3. Development of the Aviation Industry During the First World War

3.1 First Attempts until the National Flight Donation of Germany (National-Flugspende)

If we ignore purely historical experiments with airplanes, for example the "Schneider von Ulm[1]", then one can already put a beginning of the German airplane construction back into the first half of the nineties of the 18th century. The engineer Otto Lilienthal, during his gliding experiments in 1891 in Berlin-Steglitz and Lichterfelde, which continued until his fatal accident in 1896 in the hilly landscape of Rhinow near Berlin, had already installed a carbon acid engine constructed by his engineer Schauer in a glider designed as a swing-wing aircraft. In addition, he has also already sold gliders at the unit price of 500 marks, one of them to Chanute[2] to America, later Wright aviator.

Lilienthal's work "Der Vogelflug als Grundlage der Fliegerkunst", published in 1889, served the Wright brothers and many other foreign aircraft designers as a foundation for their parallel experiments. Among others, Captain Ferber[3] built his first glider in Paris in 1899 according to Lilienthal's model, the fifth apparatus according to Wright's model, and the sixth apparatus was tried out with an engine on a rotary apparatus near Nice and Brest. In the same year, Kress[4] tested his hang glider in Vienna, which had an accident in 1901.

In 1901, the Wright brothers also began their new series of experiments, and the "Aero-Club de France" organized a first annual competition for flying machines in November.

In 1902, Count Zeppelin demonstrated his propeller boat on the Wannsee near Berlin with a device for changing the transmission ratio between the engine and the propeller.

The use of observation kites for meteorological purposes also began to increase at that time. In December 1902, a Berlin registration kite reached an altitude of 4,475 meters. This was a record that was held until March 25, 1914. In the same year, at the instigation of Kaiser Wilhelm II[5], Professor Hergesell undertook kite ascents on the Baltic Sea on the S.M.S. *Sleipner*[6] on two days in June, which were continued on the same ship on the North Country voyage along the Norwegian coast.

In mid-December 1903, after the Wright brothers had completed the construction of their first powered hang glider and achieved 260 meters in 59 seconds against a wind of 36 km/h on their fourth flight, Deutsch de la Meurthe[7] donated 25,000 francs to the "Ae.C.F." in 1904 for the first plane that would perform a flight of one kilometer. But aircraft technology was not yet ready.

The following year 1905 brought across the ocean the first powered flight successes of the Wright brothers who,

Above: Wallin's rotorcraft (1906).

according to their letter of October 9 to Captain Ferber, flew 39 km in 38 minutes on October 5. However, this information was doubted because only one flight over 1,200 m is authenticated by O. Chanute and, after the founding of the Weiller Syndicate by Hart O. Berg in Paris, Wilbur Wright only achieved flights of 8 minutes 13 seconds duration in Le Mans in August 1908, while Delagrange[8] had already covered 15 km in 16 minutes in Rome. Nevertheless, a freehand drawing of the Wright apparatus in the Paris newspaper "*L'Auto*" of December 24, 1905, already shows details of the machine later demonstrated in Europe (in front a horizontal, double elevator control, in the rear a double side control, the two propulsion screws with the sideways set motor and the starting device by means of a driving rail. The prone position of the pilot in the machine was not abandoned until 1907).

For the rest, the designers were still groping for the basic system: Wallin, for example, built the first man-carrying swing-wing glider, while the Dufaux brothers[9] in Geneva built a combination of screw and hang glider. Cody[10] brought a man-carrying hang glider to 800 m altitude in Adlershof.

The first free flight in Europe was achieved in 1906 on December 12 by Ellehammer[11] over a distance of 30 to 40 m at an altitude of ½ to ¾ m on the Danish Island of Lindholm. Deutsch de la Meurthe then repeated his above-mentioned 1904 competition by donating, together with Archdeacon[12], the "Grand Prix de l'Aviation" of 50,000 francs on October 1, 1906, for the first flying machine to cover a closed trajectory of one kilometer. But even then, aircraft technology was not yet

Above: Etrich-Rumpler-Taube with Argus engine (1910).

ready to meet these conditions, even in France, and it took another year and a half to fulfill them, in addition to the previous two years from 1904 to 1906. It was not until January 1908 that the prize was won by Farman[13] on a Voisin biplane, after initial tests with lower performances had been achieved in the meantime with the hang gliders of Santos Dumont[14] (50 hp Antoinette engine, October 1906: 220 m flight distance), Bleriot[15], Voisin[16], Esnault-Pelterie[17], while Bréguet[18] made extensive tests with a screw glider at the end of 1907.

These and other names of French aircraft designers appeared in the following years together with others as aircraft industrialists, who acquired one world record after another and also exported many aircraft, including to Germany.

Here, the industry with greater means at that time was concerned almost solely with the problem of motorized airships ("lighter than air"), while the German industry still treated the problem of actual aircraft ("heavier than air") quite stepmotherly in terms of allocation and investment of funds. Since the same principle (in contrast to the French) was followed by the German military authorities, the German aircraft designers were completely dependent on their private funds to carry out their experiments.

The underdevelopment of Germany in the field of aviation was already known at that time, as the following note of the *"Deutsche Zeitschrift für Luftschiffahrt"* of March 24, 1909 shows:

"France is doing everything it can to make life as pleasant as possible for flight technicians and thus to open up safe paths for aviation. Everywhere, efforts are being made to establish favorable airfields, the main condition for testing new apparatus and training good flight technicians. In England, too, the value of good airfields has been recognized and efforts are being made to build such fields, and recently, even in Germany, efforts are being made to create a good training ground with the necessary workshops for flight technicians. Hopefully, this will also steer aviation technology in our country in a more favorable direction.

The same source reported on May 5, 1909, about a second meeting of the Verein deutscher Flugtechniker[19] (later

Right: The first attempt at Automobil und Aviatik GmbH was a disaster (1910).

Development of the Aviation Industry During the First World War

Left: Orville Wright's biplane, on which he demonstrated the first round-trip flights performed in Germany in September 1909.

Reichsflugverein) on March 17, chaired by Dr. Fr. Huth, among others, about the approval of the proposal to purchase a glider for 400 to 600 Marks, emphasizing the difficulty of acquiring a suitable engine, partly the costs to be estimated for a 30 hp engine of 90 kg weight in the manner of the Wright engine at 1,100 Marks for a 30 hp engine of 90 kg weight of the Wright type, and partly due to the fact that among the German factories that could be considered as donors or lenders, probably none would be found that would build the suitable type. Questionnaires distributed to determine the requirements for the engine.

One by one, promising ideas found the support of professional associations, which began to buy foreign aircraft for experimental purposes and to raise funds for experimental buildings – of course, on far too small a scale to be able to turn the initial failures into thoroughgoing successes through further grants.

This explains why the generous prize endowment of 40,000 marks from the industrialist Dr. Karl Lanz[20] in 1908 for the first airplane manufactured entirely in Germany, which, piloted by a German aviator, describes a flight in the shape of a figure eight around two points 1,000 meters apart and returns to the finish line, was not awarded to the airplane designer Grade[21] until the following fall (after a failure at the newly acquired Bork airfield), when he made his second attempt at the now deforested Johannisthal airfield.

The year 1908 brought only one foreign achievement in Germany, namely the first flight in Germany of the Dane Ellehammer with a length of 47 m on the occasion of a prize competition of the sports festival commission Kiel (June 28).

Likewise, the year 1909, apart from the aforementioned achievement of Grade, saw only foreign achievements: Armand Zipfel's[22] flight tests on Tempelhofer Feld with a Voisin apparatus at the instigation of the "Berliner Lokalanzeiger" from January 28 to February 4, Orville Wright's[23] flights on Tempelhofer Feld from September 4 on with a world record for duration with one passenger (1 hour,

Above: Dorners monoplane (1911).

35 minutes, 47 seconds) and altitude record of 75.3 m. On September 23, Latham[24] began his demonstration flights at Tempelhofer Feld at the request of John Rozendaal and W. Wertheim.

In August 1909, the aviation world met in France for the first major internationally attended flying competition for "heavier-than-air" flying machines, the Grande Semaine d'Aviation de la Champagne, held on the occasion of the opening of an airfield near Bétheny north of Reims and under the patronage of the President of the French Republic. Every day, up to 200,000 paying visitors flocked to the airfield, onlookers attracted by the new sensation of powered flight, as well as trade visitors from France and abroad, designers, aviators, correspondents, representatives of aeronautical and aviation associations, military officers, and politicians. The event provided an overview of the current state of performance of "heavier-than-air" flight technology, the leading designers and the best pilots had come to Bétheny with few exceptions. The entry list included no less than 38 aircraft, 23 of which actually took to the air.

A number of new records were set during the flying week, with over 120 take-offs, 87 flights of five kilometres and at least seven of more than 100 kilometres. In various competitions, the 22 participating pilots competed for high prize money, totaling about 200,000 francs, donated by the champagne wineries of the region. The winner of the Prix du Tour de Piste reached a speed of 76.9 km/h on a circuit of ten kilometres, the Prix de l'Altitude went to Hubert

Right: Pega-Emich monoplane 1913.

Development of the Aviation Industry During the First World War

Above: Gustav Otto biplane (1910/11).

Latham, who had climbed to a height of 155 meters, and the prize for the longest flight was won by Henri Farman with a distance of 180 kilometres. By flying with two passengers, Farman also won the Prix des Passagers, the prize for an aerodrome flight with the largest number of people on board.

What happened in Germany?

From September 26 to October 3, 1909, the airfield company in Johannisthal held its first flight week, during which Latham made the first cross-country flight in Germany from Tempelhofer Felde to Johannisthal, without any other notable achievements.

On September 30, Orville Wright set a world altitude record of 300 meters at Bornstedter Feld in the presence of Emperor Wilhelm I, and flew with the Crown Prince on October 22. On October 13, Captain Engelhardt[25], a student of Orville Wright, made his first flight.

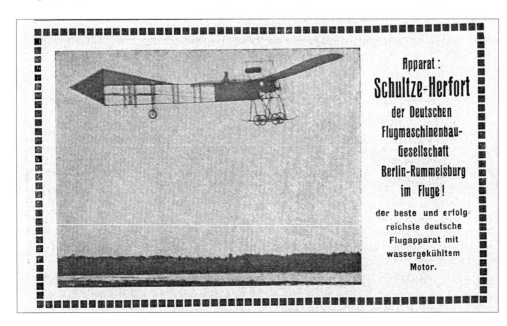

Left: Advertisement Schulze-Herford monoplane 1910.

Above: Extract from *Berliner Zeitung*, June 10, 1911.

The Cologne Aviation Week from September 30 to October 6 ended, as did the Frankfurt one from October 3 to 11, without any notable achievements; the latter, however, brought the first appearance of August Euler[26] as an aircraft pilot, who had already founded an aircraft company in 1907 and opened an airfield near Darmstadt (Griesheim) in 1908, and who also appeared at the International Airship Exhibition (Ila) in Frankfurt a. M. in 1909 as Voisin's representative and licensee.

In Berlin, the Flugmaschine Wright GmbH was founded on May 30, 1909 (later changed to L.F.G.), while other beginnings of the German aircraft industry, which started on a very small scale in 1908/09, only became known later with incipient successes.

These include, in addition to those already mentioned, the workshops and individuals of the following companies and entrepreneurs listed in the literature of the time **1909**:

Deutsche Flugmaschinen GmbH, Rummelsburg, Ing. Joachimczyk

Dorner-Flugzeugwerke, Berlin-Treptow (later Fa. Götze)

Harlan-Werke, Johannisthal

Haefelin-Flugzeugwerke, Johannisthal

Starke und Tarabochia, Darmstadt (Factory for aircraft, engines, propellers and landing gear)

Hanuschke, Johannisthal

Maschinenbauanstalt Mordhorst, Kiel

Flugapparate-Bauanstalt „Deutschland" GmbH, Berlin-Schöneberg, (Aircraft and spare parts)

Euler, Frankfurt a.M., bought the license Voisin for Germany and built on this basis his biplanes

Fliegerfabrik Hans Grade, Bork i. M.

Aero GmbH, Straßburg i.E., owned by E.E. Mathis (Sailes monopol „Antoinette" and Farman for Germany)

Pega und Emich Flugmaschinenbau, Frankfurt a.M. – Griesheim

Gesellschaft zum Bau von Luftfahrzeugen, Ing. Rumpler; Berlin

Albatroswerke GmbH, Johannisthal (initially dealt only with repairs)

Flugmaschinen- und Motoren GmbH, Berlin (Biplanes designed by Dr. Huth, Flieger Eyring).

Attempts made at that time by large industrial companies, such as Siemens-Schuckert in Bornstedt, as well as by smaller private companies, such as the yacht yard owner Oertz[27] in Hamburg, were initially abandoned.

In the summer of 1910, the Chief of the General Headquarters proposed a five-year plan for the immediate establishment of an air force in order to make the aircraft available to the army as quickly as possible for reconnaissance and liaison tasks. However, the War Ministry considered this ambitious plan to be premature and hasty; it first ordered the Cologne Commission to carry out a further examination of the aircraft manufactured in Germany and successful at the National Aviation Week in October 1910 to ensure that they were suitable for military use and could be deployed. After all the tested aircraft had been assessed as "fit for war", the commission also advised the development of an army aviation system, although this fell far short of Moltke's

Development of the Aviation Industry During the First World War

Above: Kathreiner Prize: From Puchheim near Munich to Berlin in five hours: Hellmuth Hirth on his Albatros Taube in 1911.

demands in terms of material and personnel: as a first step, a small number of proven aircraft were to be purchased, for which the War Ministry provided a sum of 110,000 marks in December 1910. The next step was to encourage private industry to develop aircraft that could meet the high technical requirements of the military in terms of load-bearing capacity, operational safety and flight performance. In addition, preparatory organizational measures should be initiated to create an expanded peacetime organization of an air force.

In the technical literature of the year **1910** are listed at other aircraft manufacturing companies:

Flugmaschine Nolte GmbH, Hannover (Construction of flying machines system Nolte);

Oertz-Flugzeug G.m.b.H., Berlin;

E. Rumpler-Luftfahrzeugbau GmbH, Berlin (Originated from the „Gesellschaft zum Bau von Flugzeugen". Construction of aircraft engines and planes);

A. Symanzig, Flugzeugbau, Borszymmen[28], Ostpreußen;

Auer, Christian, Cannstadt i. Württemberg (Factory of flying machines and propellers);

Kahan und Friedlein, Aeroplan-Gesellschaft, Gießen;

„Aviatik" GmbH Mülhausen (Aircraft and engine construction);

Hoffmann, Gebrüder, Flugzeugfabrik, Offenbach a.M.;

Albatroswerke GmbH, Johannisthal, extended to aircraft construction;

D.F.G. (Deutsche Fluggesellschaft mbH), Frankfurt a.M.

Above: Euler biplane ("*Gelber Hund*" - "*Yellow Dog*") is overflying the troops at the Kaisermanöver 1911.

Right: Prince Heinrich, the brother of Kaiser Wilhelm II, has his aircraft explained to him at the ALA exhibition stand of Deutsche Flugzeugwerke. The prince was a relentless supporter of the navy and aviation. The prince obtained his pilot's license at August Euler's flight school.

(Construction of flying machines of own design);
Gustav Otto Flugmaschinenwerke GmbH., München
(Initially repair shop, later own apparatus);

Apart from the above-mentioned, there are also mentions in the technical literature of the year **1911**:
Flugzeugwerke Schultze GmbH, Burg near Magdeburg;
Stein Aeroplanbau, Teltow-Berlin;
Flugmaschinen und Motoren GmbH.; Berlin;
O. Trinks Flugzeugbau, Berlin-Teltow;
L.V.G. (Luft-Verkehrs-Gesellschaft), originated from the „Luftfahrt-Betriebsgesellschaft mbH.", Johannisthal;
Deutsche Flugzeugwerft GmbH. (Dr. Fritz Huth), Berlin, (Metal Monoplane).
Sächsische Flugzeugwerke, later D.F.W. (Deutsche Flugzeugwerke), Leipzig;
Centrale für Aviatik, later Hansa-Flugzeugwerke, Hamburg;
Gesellschaft für Flugmaschinen und Apparatebau, Köln (GEFA monoplanes);
Professor Reißner, Aachen (Monoplane „Ente")

While the above-mentioned companies were being set up, timid attempts by the German aviation industry, as it were, the principle of "heavier than air" was experiencing its world-famous triumphs on the other side of the Vosges. One record after another passed to the West. The French army administration, which had warmly embraced the new weapon, had an inventory of 32 army aircraft at the beginning of 1911 against 234 aircraft at the turn of the year. These apparatuses were subjected to extremely rigorous testing, by the standards of the time, before acceptance. In addition to a long-distance flight of 300 km with intermediate landings in difficult terrain and a long-distance flight of 300 km without intermediate landings with a payload of 300 kg, the French army administration required a climb capability with this load of 500 m in 15 minutes.

In England, as well, interest in the new sport grew rapidly and found its numerous supporters and, more importantly, the necessary financial foundation.

German capital, on the other hand, had to wait a long time before it took up the new branch with seriousness and diligence. Above all, there was a lack of recognition of the importance of the airplane both as a weapon and ultimately as a means of transportation, because there was no longer any lack of tangible and practical experience from abroad on which to build. Apart from a few generous prize donations, among which the "Kathreiner Preis[29]" (Kathreiner Prize) and the "B.-Z. Preis der Lüfte" (B.-Z. Prize of the Skies[30]) mentioned above and the supplementary prizes to the "Lanz Prize" had a special place, German aviation lacked the money tenders that had lifted French aviation out of its infancy. Big capital was almost entirely in reserve, so that aviation was completely dependent on the more or less serious and thorough experiments of individuals, and the military administration could provide only small funds for finished products and almost no funds at all for experimental purposes.

However, the extent to which the efforts of clubs and associations to promote flight technology and thus also the German aircraft industry in detail were met with the opposite of insightful promotion of these efforts on the part of bureaucratic authorities is shown, for example, by the following announcement by the well-known Lower Rhine balloonist Dr. Bamler[31]:

"The Essen section of the Lower Rhine Association can describe how in Germany the treasury participates in the promotion of aviation. For years this section has been particularly concerned with the promotion of flying machines; it has already built one such machine and has proved its usefulness by its test flights of several kilometers. A second machine is almost completed and represents a

Development of the Aviation Industry During the First World War

Above & Below: View into the exibition hall of the ALA in Berlin.

Right: Integral-Stand at the ALA in Berlin.

completely new type which, in the opinion of all experts, has the prospect of achieving great things. Since the Section could not finance these buildings entirely from its own funds, Messrs. Krupp von Bohlen und Halbach[32] and the Rhenish Coal Syndicate joined the Section as life members and assessed their one-time contributions at 3,000 and 2,000 marks, respectively. The tax authorities, who had learned of these sums from the annual report of the association, were of the opinion that these were not membership dues, but gifts, and demanded 10% gift tax, after the money had long since been spent on the construction of flying machines. The association, of course, believes itself to be fully justified in refusing to pay the gift tax, since it has not received any gifts at all. What would the German people say if the tax authorities had levied a gift tax on the Zeppelin donation at the time! Or if it had demanded a gift tax from the Rheinisch-Westfälische Motorluftschiff-Gesellschaft, which was supported by the Minister of War with a considerable sum! In this case, however, the tax authorities believe themselves to be so far in the right that they even break the house rules at the Rhein-Elbe colliery in Gelsenkirchen, where the balloons of the section are kept, and seize the balloon "Bamler" there, despite the energetic protest of the colliery management and the association's board! In Germany, this is called promotion of aviation technology!"

Despite these general difficulties, German pilots were not inferior to their foreign colleagues in either skill or daring. Special mention should be made of the achievements of engineer Hellmuth Hirth[33], who, among other things, won the Kathreiner Prize in 1911 with his flight from Munich to Berlin on a Rumpler monoplane in 5 hours 20 minutes. Other good achievements of the German pilots, whose number was not remotely comparable to that of the French, also deserve mention here. Karl Grulich[34], the designer of the Harlan monoplane, should be mentioned, who brought the endurance records in flight with two and three passengers to Germany (2 passengers 2 : 2 : 45; 3 passengers 1 : 35 : 00). Hoffmann, who, like Grulich on Harlan monoplane, took the world record with 4 passengers (0 : 32 : 39). Schendel[35] on Dorner monoplane temporarily conquered the altitude record with 1,680 m with one passenger for Germany.

Given the low level of support for the aircraft industry from big business, it is understandable that individual types of aircraft, such as those brought to market by the Haefelin company, had to have their permission to take part in competitions withdrawn by the German Aviation Association, the supreme German sports authority, because their handling was associated with too much uncertainty and danger to life. There was a lack of financial support for sufficient development of a safe design, so that failures for such young companies usually had to lead to complete cessation of operations.

The expected support did not materialize even when, at the Paris Air Show at the end of 1911, the small German aircraft industry received the greatest and undivided admiration for its achievements in the form of aircraft from the companies Albatros and Aviatik. The Albatros biplane in particular did not fail to make an impression on our Western competitors for the conquest of the air.

With the progress made in reliability and performance, flight duration and distance, stability and load-bearing capacity, the army headquarters gradually began to attach real importance to the airplane for warfare, as indicated by the War Ministry's directive to extend the artillery combat trials against airships to aircraft, with the aim of denying the enemy the use of a practically useful means of observation and attack. In September 1911, aviation units took part for the first time alongside airships in a major troop exercise, the annual Kaisermanöver.

The Chief of the General Staff was very satisfied with the performance of the airmen, and his view that the airplane had already developed into a valuable means of reconnaissance was reinforced by reports from abroad. Von Moltke considered the army's existing air force equipment to be inadequate and militarily questionable; in the event of war, he expected numerous enemy aircraft to be deployed, especially on Germany's western border.

However, the international exhibitions give the most accurate picture of the state of aircraft technology and of the aircraft industry, as has been the case since 1900 for the automotive industry, whose national and international exhibitions are held in Germany, France, England, and Italy at the same venues and by the same companies (industrialists, associations, clubs). The "Ala", the General Aircraft Exhibition in Berlin, from April 3 to 14, 1912, presented essentially the same picture with regard to the technology of the German aircraft industry: good products, some of which were at least equal to those of foreign manufacturers, quite a large number of visitors - but everywhere the cry for help; the hunt for the absolutely necessary capital, which is mostly futile in view of our rather thinly spread invitations to tender and the lack of interest on the part of the widest circles.
3_1_1l Prince Heinrich, the brother of Kaiser Wilhelm II, has his aircraft explained to him at the ALA exhibition stand of Deutsche Flugzeugwerke. The prince was a relentless supporter of the navy and aviation. The prince obtained his pilot's license at August Euler s flight school.

From the exhibitors and exhibited objects at the ALA 1912 may be mentioned:

a) Aircraft

Albatroswerke GmbH, whose biplane already caused a sensation at the Paris Aero-Salon, with military types 1911 and 1912;

Deutsche Flugmaschinen Wright GmbH, with a new "Rekord" racing biplane;

Otto-Flugzeugwerke, Munich, with racing (military) biplanes (main focus: easy dismantling) and monoplanes 1912;

Goedecker-Flugzeugwerke with military monoplane;

Rumpler-Luftfahrzeugbau GmbH with type 1912 Taube and twin-engine Loutzkoy Taube;

Flugzeugwerke August Euler, Frankfurt a.M., with biplanes and triplanes;

Harlan-Flugzeugwerke with the well-known monoplane (military type), on which Ing. Hoffmann beat the world record with four passengers with 00 : 32 : 29;

Fliegerwerke Hans Grade, Bork, which exhibited the old, small monoplane which has undergone no significant changes since winning the Lanz Prize.

Flugzeugwerke Gustav Schulze, Burg near Magdeburg, with a monoplane with a low-slung pilot's seat;

Automobil- und Aviatik-A.G., Mulhouse, which brought military biplanes (light transportability, short undercarriage, 95 km/h) and monoplanes;

Dorner-Flugzeugwerke with the well-known Dorner monoplane;

Flugzeugwerke Haefelin & Co. with armored steel airplane with 155 hp air-cooled engine;

Deutsche Flugzeugwerft GmbH, all-metal monoplane, Dr. Huth design.

b) Engine and propeller factories, etc.

Argus-Motorengesellschaft, Berlin-Reinickendorf (four-cylinder stationary engines);

Daimler-Motoren-Gesellschaft, Stuttgart, (four-cylinder stationary engines);

Rheinische Aerowerke, Düsseldorf-Oberkassel (five-cylinder rotary engine, three-cylinder radial engine);

Neue Automobil-Gesellschaft, Berlin (N.A.G.) (four- and six-cylinder stationary engines);

Arthur Delfosse , Cologne-Riehl (seven-cylinder rotary engine);

Georg Hoffmann, Frankfurt a.M. (nine-cylinder rotary engine);

Garuda-Flugzeug- und Propeller-GmbH, to whose account also the construction of the "Möwe" designed by Dr. Geest and also exhibited, brought the proven Geruda propeller.

This short list of exhibitors and exhibits at the "ALA" already shows the versatility of the German aircraft industry even before the Nationalflugspende; it shows that the industry was not only made up of aircraft factories, as they were mostly newly founded, but that a large number of already emerging companies were also involved in the manufacture and supply of stock and accessory parts for aircraft. It should be mentioned at this point that the exhibited aircraft of the Rumpler and Albatros companies were purchased by the army administration.

As far as the engine question in particular is concerned, it may be briefly summarized at this point that Ader's first

Above: With this Albatros water biplane, the Kaiserliche Marine made its first attempts at wireless communication as early as 1912. In addition to the antenna wires, various parts of a radio system can be seen.

flight in 1891 (about 250 m without touching the ground) was achieved by means of a steam engine, while Lilienthal was concerned with an aircraft carbonic acid engine specially constructed by his engineer Schauer, and only the sufficient development of light aircraft engines for liquid fuels to a sufficiently favorable ratio of weight to engine power led to the successes in aircraft construction itself. However, while the first aircraft designers, such as Wright, Grade, Antoinette, and later aircraft designers, such as Dr. Fritz Huth etc., who did not succeed in developing such engines due to a lack of sufficient funds, also began to turn their attention to aircraft engines at the time before the "Ala", even larger engine factories with many years of experience in motor vehicle and boat engine construction. It was obvious that Daimler-Motoren-Gesellschaft would soon have to play a major role in this field as well, once this vast experience in the construction of powerful and light engines was available. Other circumstances, such as the association with Automobil- und Aviatik-A.G. and with the pilot Emile Jeannin, a brother of the founder and head of Argus-Motoren-Gesellschaft, Henri Jeannin, favored, for example, the testing of this type of engine by Argus-Gesellschaft.

A powerful impulse to improve German aircraft engines was given by Kaiser Wilhelm II's endowment of a prize of 50,000 marks on January 27, 1912, his birthday. The decree read:

"For the promotion of German aviation, I intend to donate a monetary prize of 50,000 marks from my coffers, which I will award for the best German aircraft engine on my birthday next year. A committee is to be formed for the purpose of awarding the prize and for examining and evaluating the entries received. This committee is to consist of members of the Imperial Automobile Club, the Imperial Aero Club, the Association of German Motor Vehicle Industrialists, as well as one representative each from the Reich Office of the Interior, the Reich Navy Office, the War Ministry, the Ministry of Spiritual Affairs, etc., and the Technical University of Berlin. Affairs and the Technical University of Berlin. I request you to report to me on the progress of the matter and to submit by the beginning of January next year the proposal of the jury to be formed for the award of the prize.

Development of the Aviation Industry During the First World War

Berlin, January 27, 1912
Wilhelm I.R.[36]"

The Kaiser Prize for the best German aircraft engine was a competition of German engine manufacturers with a prize money of 50,000 marks. Various authorities of the German Reich and the Prussian state increased the prize money to a total of 125,000 marks.

With this appeal to German engine manufacturers, for the first time the elites of the Empire, represented by the Kaiser himself as well as his leading authorities, actively contributed to the development of military aviation. Unlike the Zeppelin donation, where the German people raised more than 6 million marks within a short time for the construction of Zeppelin airships, the state became an active part of aviation, here especially for the development of the German aircraft industry.

This call to all manufacturers of engines was not only intended to make up for the technological shortfall that had occurred in the meantime, especially with France. One reason was to get the engine manufacturers interested in aviation in the first place. In the situation at the time, a handful of German airships were powered by their own engines. Only about half of the engines of the less than 200 airplanes existing in Germany were equipped with domestic engines, which, moreover, were distributed among 5-7 manufacturers.

In other words, it was not at all interesting for the engine industry to invest effort and money in aircraft engines. After all, there was enough sales potential in the burgeoning automobile and shipbuilding industries.

The applications for the competition were received at the beginning of July, and it was only now that it became clear how extensive the preparations had to be. The imperial wake-up call and the amount of the cash prizes had achieved the purpose of calling Germany's relevant industry to the scene to a surprising extent. No fewer than 41 different types of engines were entered by 23 applicants. A total of 65 engines took part in the competition. The winner was the four-cylinder Benz FX engine from Benz & Cie, Rheinische Automobil- und Motorenfabrik AG, Mannheim.

The competition for the best German aircraft engine had just been initiated with promising results, when the German people were now called upon to promote national defence through a strengthened aviation industry.

Endnotes

1 Berblinger, swingwing glider 1811.
2 Octave Chanute (February 18, 1832, Paris – November 23, 1910, Chicago, Illinois) was an American Engineer and aviation pioneer. He provided many budding enthusiasts, including the Wright brothers, with help and advice, and helped to publicize their flying experiments. At his death he was hailed as the father of aviation and the heavier-than-air flying machine.
3 Ferdinand Ferber (* January 8, 1862 in Lyon; † September 22, 1909 in Beuvrequen) was a French officer (Capitaine, Eng. Captain) and aviation pioneer. He used the pseudonym De Rue in flying competitions.
4 Wilhelm Kress (also Wilhelm Kreß) (* July 29, 1836 in St. Petersburg; † February 24, 1913 in Vienna) was an Austrian aviation pioneer and designer. He built a powered aircraft at Lake Wienerwald west of Vienna in 1901, however the seaplane capsized during test flights without ever taking off.
5 Wilhelm II, full name Friedrich Wilhelm Viktor Albert of Prussia, (b. January 27, 1859 in Berlin; † June 4, 1941 in Doorn, Netherlands) from the House of Hohenzollern, was the last German Emperor and King of Prussia from 1888 to 1918.
6 SMS *Sleipner* (ex torpedo boat S 97) was a dispatch boat at the disposal of the German Emperor Wilhelm II. It was usually used together with the imperial yacht *Hohenzollern*. Due to its function such as transporting state guests, the "*Sleipner*" was one of the most famous warships of the Imperial Navy.
7 Henry Deutsch de la Meurthe (* September 25, 1846 in La Villette (Paris); † November 24, 1919 in Ecquevilly, Yvelines department), born as Salomon Henry Deutsch, was a French industrialist and patron of the arts.
8 Ferdinand Marie Léon Delagrange, also Léon Noël Delagrange (* March 13, 1872, Orléans, France; † January 4, 1910, Bordeaux, France). In 1907, he was one of the first aviation pioneers to order a biplane from the Voisin brothers, who, with the implementation of this order, the Voisin-Delagrange I, created the prototype of their successful Voisin Standard model and thus established themselves as a manufacturer of aircraft.
9 Armand Dufaux (* January 13, 1883, Paris; † July 17, 1941, Geneva) and Henri (b. September 18, 1879, Chens-sur-Léman, France; † December 25, 1980, Geneva, Switzerland), French-Swiss aviation pioneers, inventors and designers, constructed a rotorcraft that was patented on February 24, 1904, and first publicly demonstrated on April 14, 1905.
10 Samuel Franklin Cody (* March 6, 1867 in Davenport, Iowa; † August 7, 1913 over Laffan's Plain, Aldershot, southern England, in a plane crash) was an English aviation pioneer.
11 Jacob Christian Hansen Ellehammer (* June 14, 1871 at Bakkebølle in Sydsjælland, Denmark; † May 20, 1946 at Gentofte) was a Danish aviation pioneer.
12 Ernest Archdeacon (* 1863 in Paris; † 1950 in Versailles) was a prominent French lawyer of Irish descent and an automobile and aviation pioneer.
13 Henri Farman (b. May 26, 1874 in Paris; † July 18, 1958 ibid) was a French track cyclist, aviation pioneer and entrepreneur.
14 Alberto Santos Dumont (* July 20, 1873 on the Fazenda

Cabangu near Palmira in the Brazilian state of Minas Gerais; † July 23, 1932 in Guarujá/São Paulo state) was a Brazilian airship pilot, powered flight pioneer and inventor who helped shape the beginning of motorized aviation, especially in his country of creation, France. After several flights in various self-built airships, he also performed the world's first public powered flight in an airplane in 1906.

15 Louis Charles Joseph Blériot (* July 1, 1872 in Cambrai; † August 2, 1936 in Paris) was a French aviation pioneer. In the Blériot XI, he became the first person to cross the English Channel in an airplane on July 25, 1909. His flight from Calais to Dover lasted 37 minutes at an average altitude of 100 meters.

16 Gabriel Voisin (* February 5, 1880 in Belleville; † December 25, 1973 in Ozenay) was a French aircraft and automobile designer and the vice director of the Aéro-Club de France.

17 Robert Albert Charles Esnault-Pelterie (* November 8, 1881 in Paris; † December 6, 1957 there) was a French aeronautical and rocket pioneer.

18 Louis Charles Breguet (* January 2, 1880 in Paris; † May 4, 1955 in Saint-Germain-en-Laye) was a French aircraft designer and a co-founder of the airline Air France.

19 The "Association of German Aeronautical Engineers" was founded in Berlin on February 10, 1909. The initiators of the association are Fritz Huth (1872-1948), the airship constructor August von Parseval (1861-1942), the aviation pioneer and later State Secretary at the Reichsluftamt August Euler (1868-1957) and the engineer and editor of the magazine "*Flugsport*", which has been published since 1908, Carl Oskar Ursinus (1878-1952), who gather around them a number of serious aviation technicians. According to the statutes, the aim of the association is "to promote the construction of flying machines and the practice of aviation". Fritz Huth, the owner of the "Flugmaschinen- und Motorengesellschaft mbH" in Berlin-Johannisthal, is appointed chairman of the association.[1] In March 1910, the association was re-established in Frankfurt am Main as the "Verein Deutscher Flugzeug-Industrieller" (Association of German Aircraft Industrialists), thus creating the first German aviation industry association. The founding companies joining the association include Albatroswerke GmbH, Automobil- und Aviatik GmbH, Dorner Flugzeug-Gesellschaft mbH, August Euler Flugzeugwerke, Hans-Grade Flieger-Werke, Harlan Flugzeugwerke GmbH, Edmund Rumpler Luftfahrzeugbau and Flugmaschine Wright GmbH. In 1919, the company was renamed the "Verband Deutscher Luftfahrtzeug-Industrieller" (Association of German Aircraft Manufacturers) and in the 1920s the "Reichsverband der Deutschen Luftfahrtindustrie" (Imperial Association of the German Aviation Industry).

20 Karl Lanz (* May 18, 1873 in Mannheim; † August 18, 1921 ibid; full name: Karl Wilhelm Konstantin Philipp Lanz) was a German mechanical engineer, entrepreneur and promoter of technical innovations. On April 22, 1909, Lanz founded the company Luftschiffbau Lanz & Schütte GmbH in Mannheim-Rheinau together with Johann Schütte. It produced 22 rigid airships in the years to come and also aircraft for the military during World War I, especially at the Zeesen site. Lanz's role was limited to that of investor. He had no influence on technical developments.

21 Johannes Gustav Paul "Hans" Grade (* May 17, 1879 in Köslin, Pomerania Province, Prussia, now Poland; † October 22, 1946 in Borkheide) was a German mechanical engineer, entrepreneur and aviation pioneer.

22 Armand Zipfel (1883-1954), French aviation pioneer and aircraft builder.

23 Orville Wright (born August 19, 1871 in Dayton, Ohio; † January 30, 1948, same place) was a American airplane builder and pilot.

24 Hubert Latham (* January 10, 1883 in Paris; † June 25, 1912 at Fort Archambault, now Sarh) was a French aviation pioneer. He set a number of flight records in 1909 and 1910 (1910, altitude record of 1384 meters) and first flight achievements. On September 27, 1909, he made the first cross-country flight in Germany in an Antoinette monoplane from Tempelhofer Feld to Johannisthal airfield.

25 Paul Emil Engelhard (auch Paul Engelhardt; * 27. Juli 1868 in Münster; † 29. September 1911 in Berlin-Johannisthal) war ein deutscher Flugpionier.

26 August Euler (* November 20, 1868 in Oelde; † July 1, 1957 in Feldberg (Black Forest); born August Heinrich Reith) was a German engineer and manager who gained notoriety as the first holder of a German "airplane pilot's license" and aviation pioneer, and after World War I headed the Reichsluftamt or Reichsamt für Luft- und Kraftfahrwesen at the Reichsverkehrsministerium for a few years.

27 Max Oertz (* April 20, 1871 in Neustadt in Holstein; † November 24, 1929 in Hamburg) is one of the great German yacht designers. He had a decisive influence on German yacht building at the beginning of the 20th century.

28 Borzymy (German Borszymmen, 1936-1938 Borschymmen, 1938-1945 Borschimmen) is a village belonging to the municipality of Kalinowo (Kallinowen, 1938-1945 Dreimühlen) in eastern Masuria in the Polish Warmia-Masuria Voivodeship (Lyck District).

29 This competition was organized 1911 by the coffee company "Kathreiners" for the first successful flight from Munich to Berlin within 36 hours.

30 Preis der Lüfte 1911" was launched from Johannisthal airfield near Berlin on June 11, 1911. Around 450,000 marks were available as prizes, of which around 350,000 marks were raised by the participating cities and 100,000 marks by the newspaper Berliner Zeitung B.Z.

31 Karl Bernhard Bamler (* October 29, 1865 in Cammin; † March 27, 1926 in Essen-Rellinghausen) was a German meteorologist, teacher and pioneer of free ballooning.
32 Gustav Georg Friedrich Maria Krupp von Bohlen und Halbach (* August 7, 1870 in The Hague, Netherlands; † January 16, 1950 in Schloss Blühnbach) was a German diplomat and later, after marrying Bertha Krupp, chairman of the supervisory board of Friedrich Krupp AG.
33 Hellmuth Hirth (* April 24, 1886 in Heilbronn; † July 1, 1938 in Karlsbad) was a German aviation pioneer, aircraft and aircraft engine designer.
34 Dr. Ing. Karl Grulich (* 03.09.1881 Halle a. S., † 29.07.1949 Bad König) worked as a design engineer at the Gothaer Waggonfabrik, among others.
35 Schendel was born in 1885 and was a shipbuilding engineer by profession. He learned to fly in November 1910, took his pilot's exam in March 1911 in Munich - Buchheim and then joined Dorner Werke as a flight instructor and engineer. On June 9, 1991, he and his mechanic crashed fatally in Johannisthal.
36 "Wilhelm I.R." The abbreviation stands for "Imperator Rex", i.e. "Emperor and King" - since the founding of the German Empire in 1871, the customary name suffix of German monarchs.

After National Flugspende

3.2 From the Nationalflugspende to the Beginning of the War (Spring 1912 to Summer 1914)

Dr. Count Posadowsky-Wehner, President of the Board of Trustees of the Nationalflugspende (National Flying Donation), was at the head of the men who, under the protectorate and at the suggestion of Prince Heinrich of Prussia, who had been of great service to German aviation, issued an appeal for the formation of a (second) National Flying Donation on April 21, 1912 - as the only way to realize the wish of the German nation to catch up with or, if possible, surpass the enormous lead of other countries, especially France, in aviation.

August Euler's last German endurance flight record of 3 hours 6 minutes flying time of October 25, 1910 had been beaten more than three times by the French world record of 11 hours. There was only one aircraft industry for foreign army administrations: the French. The reasons, especially the lack of large capital investment without guarantee of profit, have been discussed in the concluding part of the previous chapter, so the only way out was to call upon the patriotic sense of sacrifice of the German people, as had already been done after the destruction of the Zeppelin LZ 4 near Echterdingen.

The situation around aviation was as shown below.

The cost of air armament was (see Groehler "Geschichte des Luftkriegs"), converted into marks:

The wording of the appeal of the German Reich Committee of the National Flight Donation was as follows:

"With pride we Germans may call the man ours who first realized the longing of centuries: Zeppelin. However, the rapid development of aviation with the advent of the airplane forces us to make the utmost effort to avoid being pushed behind by the sacrifice and energy of other nations. If anywhere, it must always be said here: "Germans to the front!" It is not playful ambition that demands this of us, but here it is a matter of preserving our fame as the first Masters of Applied Science, here it is a matter of creating values which shall secure for us Germans a place of honour in the history of all times.

Not everyone has the privilege of putting his physical and mental strength personally into the service of this honourable national task. But everyone can contribute a mite, so that the

Aircraft Stocks in 1911

Country	Army	Navy	Civil	Total
France	161	10	400	571
Great Britain	57	31	167	255
Russia	23	5	41	69
Germany	50	2	101	153
Italy	31	4	36	71
China	1	-	1	2
Japan	4	4	3	11
US	11	3	301	315

Source: *Groehler Geschichte des Luftkriegs*

Aircraft Stocks in 1912

Country	Army	Navy	Civil	Total
France	259	1	422	682
Great Britain	23	6	130	159
Russia	99	1	50	159
Germany	46	2	100	148
Italy	22	4	50	76
China	1	-	2	3
Japan	10	4	2	16
US	3	2	~750	~755

Source: *Groehler Geschichte des Luftkriegs*

Development of the Aviation Industry During the First World War

Above: August Euler advertises his airplanes after breaking the hour record.

Above: The magazine *"Flugsport"* published a bi-weekly overview of aviation world records.

Cost of Air Armament				
Year		Germany	France	Great Britain
1910	Total	3,918,400	–	–
	Airforce	307,500	4,100,000	2,100,000
1911	Total	4,653,500	–	–
	Airforce	2,359,100	12,200,000	2,600,000
1912	Total	10,098,550	–	–
	Airforce	3,959,150	26,000,000	4,600,000
1913	Total	39,985,650	–	–
	Airforce	17,475,100	32,900,000	10,500,000
1914	Total	62,759,350	–	–
	Airforce	27,029,600	37,300,000	20,000,000
Source: *Groehler Geschichte des Luftkriegs* converted into Marks.				

After National Flugspende

Above: Postcard. The revenue from the sale was part of the National-Flugspende.

total will of the German nation is the powerful engine that helps the German flying machine to victory.

This is a national flight donation, in which no one wants to miss out, nor should they, a national donation for German aviation and German aviators.

For the men who, as pioneers of a new, great cultural task, are dedicating their lives in the patriotic endeavour to secure for Germany an equal place in the competition between nations in this field as well, deserve the working support of the entire nation.

Above all, however, the Nationalflugspende is intended to provide the means to continue working unceasingly on the perfection of flying machines, on the training of pilots, so that the dangers are reduced, and the performance increased. With the help of the national aviation donation, the development of a technology should also be promoted, which will open up ever new fields of work and employment. Frequent and extensive competitions are intended to stimulate inventiveness, courage and energy, and to produce ever more proud achievements by men and machines. In

Above: The appeal for the National Flight Donation 1912.

short, today's airplane, whether it circles vigilantly in the skies in the hour of national danger or whether it hurries through the land in peaceful competition between nations as the newest means of modern transportation and as a winged messenger of patriotic efficiency, should be ready and able at any moment to fulfill what the dictates of the hour demand of it.

When, after the disaster in Echterdingen, a stormy national enthusiasm swept through the German people and millions were raised in a few weeks, the millions alone were not the tangible gain of the national enthusiasm. The fact that the German people unanimously rose to a national deed showed the world the elementary force of the German people's will.

For the people - by the people!
Signed by
Heinrich, Prinz v. Preußen, Protektor
Dr. Graf v. Posadowsky-Wehner, Präsident des Komitees
Franz v. Mendelssohn, Banker, Treasurer

This long-awaited appeal, already announced at the "Ala" in Berlin on April 3, 1912 by the emperor's brother, was published on April 12, 1912.

If the German people had known at the beginning of 1912 that the Fliegerwaffe would play such an eminently

Above: & Above Right: Summary of the aircraft purchased from the funds of the National Flugspende.

important role in the First World War, the success could hardly have been greater, for within a few months 7 million marks had been subscribed at the disposal of the Board of Trustees. In addition to the above-mentioned president of the board, the member of the board, Ministerialdirektor Dr. Lewald[2], the managing trustee, Geheimer Oberregierungsrat Albert, and his deputy, Amtsrichter Dr. Trautmann, ensured that not only the sporting associations but also the aircraft industry itself were heard with regard to the proposals for use. The main task in this direction was the creation of as many military service pilots as possible, but also one-year volunteers with good minimum performance as a substitute for the active flying corps for the army administration in case of emergency and, on the other hand, as high an average level of aircraft performance as possible, for which purpose in particular the system of hourly flight bonuses, the training of appropriately trained student pilots at the expense of the national flight donation and participation in prize competitions were created on a large scale.

The Nationalflugspende can therefore be understood as a direct support measure for industry, since the increased need for aircraft was foreseeable. Germany's military administration stated bluntly that the aircraft would play an active role in the armed forces as soon as sufficient pilots were available.

On the other hand, in the course of the Nationalflugspende, new aeronautical challenges were constantly being defined, which spurred the industry to pay increased attention above all to the reliability of the flying machines.

As a program for the use of the Nationalflugspende was established shortly after the above appeal:

"The aim is to perfect flying machines, to train pilots, and to care for the surviving dependents of those men who have sacrificed their lives for this great cause; furthermore, the support of competitions, both the flying competitions and the engine competitions, and finally the technical and scientific penetration and development of aviation…"

Since the German army administration still believed that it had to direct its main attention to the "lighter than air" system, i.e. in particular the expansion of the Zeppelin airship and the production of such, it becomes understandable when, for example, Captain (ret.) Hildebrand[3] wrote in the „Deutsche Luftfahrer-Zeitschrift" (D.Z.L.) on June 26, 1912:

"It is generally known in professional circles that our aircraft industry is in dire straits, but until recently hardly any serious effort was made to provide adequate support for this Cinderella of technology. For this reason, the Airmen's Association has devoted a particularly large amount of space to the industry in its memorandum sent to the authorities and legislative bodies. It is most regrettable that we, in whose fatherland aviation technology was born, have had to lag so far behind other countries. Otto Lilienthal is acknowledged to be the first pioneer of flight. However, since his death, from 1896 to 1908, aircraft "heavier than air" have been completely neglected in our country. The few people who foresaw the future of kites, such as Government Councillor J. Hofmann, were ridiculed, and only with difficulty and hardship were they able to obtain money to hire their experiments. Since the army administration was very reluctant at first, the industry did not see any acceptable sales area in front of it, so only a few could decide to put money into this seemingly lost cause. However, as soon as the army administration began to place orders, the number of factories immediately grew considerably, as it was generally believed that the need would come to an end. For the time being, however, the aircraft industry is still in the doldrums. The factories have spent a great deal of money on trials, but they are earning little or nothing, and most are even putting on weight. The War Department has recently issued a warning against the establishment of new factories, a warning which is entirely justified.

The army administration does not yet have the necessary funds to place really large orders, and the sport is still too much in decline. Only a few factories have actually been able to make a surplus by winning larger prizes in the past year, but they have all suffered in terms of production. The prizes that can be given at the individual events are extraordinarily small in view of the large expenses, and only the first, and perhaps sometimes the second, winner can earn something or at least cover his expenses..."

Detailed information on the use of the Nationalflugspende in Germany was published nine months later in an essay by Oberleutnant Mickel in the "Deutsche Luftfahrer-Zeitschrift" (March 19, 1913), which shall be partly reproduced here:

"In contrast to other countries, where the national aviation donation has been used entirely, or at least to a large extent, for the purchase of aircraft for the army and navy, we have taken a more far-sighted approach. A one-time mass order of airplanes may provide employment and earnings for the industry for some time, but it does little to promote its further development. As long as the industry is fully occupied with the construction of the ordered type, it hardly has time to bother with tests for improvements and new designs. Once the ordered machines have been delivered, however, the brief favourable economic situation has led to so many new companies being founded that the competition between the firms for the smaller orders now expected has only increased, and the prospects of the individual have thus become even worse. A healthy aircraft industry can only be brought into being if it is protected from too great fluctuations in employment. It is the task of the military and naval administrations, which have the greatest interest in the development of an efficient aircraft industry and which, as the main customers, are in a position to do so on their own, to ensure this…

Looking over all the measures of the National Flight Donation, one cannot but admit that indeed the most important areas have been taken into account. Certainly, there is still a lot to be done. … It was self-evident that many hopes could not be fulfilled. But one may expect that the enormous work that had to be done will be worthwhile, and that, just as the Zeppelin donation once laid the foundation for Germany's present supremacy in airship construction, the National Flight Donation thus applied will contribute to bringing German aviation to the place where it belongs in the world. … In this way, it may have helped to make true the prophecy of the French General Bonal: "Germany, in adopting aviation for the army, will bring method into it as befits her national genius, while France mostly neglects it. This will create for Germany a superiority which will outweigh what France owes to the liveness and vitality of its temperament."

Returning to the development of the German aircraft industry from the time of the appeal for the National Flight Donation at the end of April 1914, the elimination of the described emergency situation of the German aircraft industry was naturally only possible very slowly and step by step, just as the raising of the 7 million Mark National Flight Donation and the subsequent described decisions on the use of these funds only took place gradually over a large period of time after enormous individual propaganda.

And the national flight donation helped and did not fail to achieve its goals!

The great moral effect alone had to spur on the German designers and aircraft manufacturers, all the more so since at the same time many a good prize was offered to them by the usual competitions. As far as the German air shows were concerned, the year 1913 provided quite lively numerous highlights: In Berlin, in addition to a flying week at Johannisthal, a repeat of the "Rund um Berlin" flight, which had already taken place in 1912, was planned for the end of August. Other major flying meetings were planned by the West German Airline in Gelsenkirchen in August, by the North German Airship Association in Kiel, and by the Silesian Association for a flying meeting in Wroclaw in mid-June. More numerous were the cross-country flight projects. The Central German Association planned a sightseeing flight through Central Germany, the Northwest Group a sightseeing flight through Northwest Germany, the East German Group a sightseeing flight through East Germany, the Reichsflug-Verein a north-south flight from the sea to Munich, and the Southwest Group the Upper Rhine flight from May 10 to 19; this was to be followed by a seaplane competition on Lake Constance. Another seaplane competition was projected by the Frankfurt Flugsport-Klub in Heiligendamm.

The last-named Heiligendamm hydroplane meeting was

Development of the Aviation Industry During the First World War

Above: Viktor Stoeffler with an LVG DD biplane.

Above: Viktor Stoeffler (here in an LVG monoplane) was awarded by the National-Flugspende.

particularly richly endowed. The only serious competitors were Rumpler, Albatros and Aviatik. In contrast, it is significant that foreign competitions, such as the seaplane competitions at St. Malo (France) and Tamise (Belgium), which were held around the same time, had incomparably more lively participation and results with smaller tenders (Heiligendamm 100,000 Mark total prizes), which were also far above those of the German competitions. At that time, seaplanes were required to have wheels and float skids at the same time, the former to be retractable, if possible, but this was abandoned after the results of the competitions.

The greatest and most lasting incentive, however, came from the large individual prizes awarded by the Nationalflugspende, which resulted in a general arms race in German aviation. With tenacious energy, German aviators and aircraft factories initially set about bringing the world altitude record to Germany.

But what the Germans wanted to see, the successful

Cost of Air Armament				
Award Winner	Pilot	Distance (km)	Prize (Marks)	Date
Aviatik, Mülhausen Aviatik DD, 100 PS Mercedes	Viktor Stoeffler	2,079	100,000	October 14, 1913
Waggonfabrik Gotha Gotha-Taube, 100 PS Mercedes	Ernst Schlegel[6]	1,497	60,000	October 22, 1913
Waggonfabrik Gotha Gotha-Taube, 100 PS Mercedes	Karl Caspar[7]	1,381	50,000	October 17, 1913
Albatros, Johannisthal Albatros-DD, 100 PS Mercedes	Robert Thelen[8]	1,373	40,000	October 14, 1913
Militärverwaltung Albatros, Mercedes 100 PS	Oltn. Kastner	1,228	25,000	October 27, 1913
Militärverwaltung Aviatik DD, 100 PS Mercedes	Lt. Geyer	1,173	15,000	October 27, 1913
Jeannin, Johannisthal Jeannin-Stahltaube, 100 PS Argus	Otto Stiefvater	1,170	10,000	September 17, 1913

Right: Johannisthal 1913. In the top left corner, spectator stands can be seen next to the two airship hangars. The existing industry had settled in various small sheds in the south-west of the site (not visible in the photo).

combat of the most important records not only still held by the French, but in the meantime considerably improved, had not yet been achieved. In particular, it was the magnificent achievements of French pilots in completing long distances, such as Brindejonc's[4] flight from Paris to Warsaw, and the long-distance flights of other French aviators, admired throughout the world, that convinced the Board of Trustees of the National Flight Donation of the need to counter foreign superiority as quickly as possible.

After consultations between the Board of Trustees on the one hand and the representatives of the German army administration and navy with the involvement of the aircraft industry on the other, it was concluded, after listening to the manufacturers in detail, that it would not be expedient to further postpone the long-planned invitation to tender for the large long-distance flights. Rather, it was felt that it was urgently necessary not to allow the belief in the inferiority of German aviation to arise in the world. It was therefore necessary to follow the incentive given by the superior French performance and to take advantage of the moment by inviting tenders with high prices.

The invitation to tender for the prize was quite generous. The minimum performance required was to fly a distance of 1,000 km within one day. A total of 200,000 M was offered. The amount of the prizes was necessary because it was not only a matter of achieving higher performance by the pilots, but also a competition for German industry. The performance of German aircraft designs and types was to be demonstrated to the world. The industry was to be effectively stimulated to make technical improvements.

The tenders had a success that far exceeded even high hopes. Flights that had been considered an impossibility not only in Germany only a short time ago were carried out. The world record was beaten by Viktor Stoeffler[5] with a flight of 2,079 km, which he covered in 24 hours. The most important world record - the record in cross-country flight - had become German. Apart from the greatness of the distances, it was especially recognized abroad without envy that for the first time flights at night time over long distances were carried out with such substantial success. The following small table shows that Stoeffler's achievement was not achieved in isolation, but that the prize money of 300,000 M alone was shared by seven aviators.

However, the Nationalflugspende was not content with the achievement of this one, albeit important, world record and offered high premiums for the achievement of the other most important world records. Here, too, the success was a great one. On July 11, 1914, the aviator Reinhold Böhm beat the French world record in endurance flight and increased it to 24 hours 12 minutes. It should be borne in mind that no mechanical means of transport has yet been in operation for anywhere near such a time without the addition of new fuel...

Thus, thanks to the Nationalflugspende, Germany is in possession of the most important world records, the record in cross-country flight, in endurance flight and in high-altitude flight. The goal which the National Flight Donation had to strive for according to the will of the German people was thus essentially achieved.

These achievements had put German aviation, and with it the German aircraft industry, in the first line. Even the French sportsmen could not deny the success and even honestly expressed their admiration for what had been achieved in such a surprisingly short time. The well-known magazine "*L'Auto*"[9], Paris, wrote on October 15, 1913:

"It seems to be a done deal that the aviators from across the Rhine want to triumph over us. Without drums and trumpets one of their best pilots has covered 2000 km. Surely this huge long-distance flight shows that German aviation has overcome the period of fumbling. I do not yet know the wonderful apparatus - a biplane - nor the engine that worked

Development of the Aviation Industry During the First World War

Above: Naval Air Station 1913/14.

in it, which allowed to accomplish such feats. But one can only bow to the success of the industry, which unfortunately is not ours."

The journal *Aero*[10] wrote:

"After a long, very long period of groping, flight technology in Germany seemed to have passed into the practical period. Only 8 months ago, German officers and private pilots barely dared to make 300 to 400 km flights. Today Germany can reap where it has sown. The blood of heroes has fertilized the soil. Long distance air journeys have been accomplished; people fly there at night. The pilots are bold and in the face of our virtuosos they perform heroic deeds. What causes this zeal of the German pilots? In our opinion, it is quite natural; the prize of national aviation drives the individual pilots to outdo each other in boldness. The flying technique among our neighbours is encouraged by pith pieces. The long air journeys multiply in Germany, and soon the Reich will not be large enough for the journeys of the birds beyond the Rhine (flight Friedrich and Reichelt to Paris, Stoeffler from Johannisthal to Warsaw, Hirth to the upper Italian lakes). Yesterday, a German aviator achieved a wonderful feat: Stoeffler's flight of 2,000 km. This is a record, and it is fitting to congratulate the person who accomplished it. By this wonderful journey, the German pilot places himself among the best flying men on the other side of the Rhine, and he can only be congratulated."

This recognition of German flight achievements shortly before the beginning of the First War was all that more significant because in the previous year, 1913, the French had actually been able to demonstrate successes and distances in cross-country flight with which the top German achievements at the time could not even remotely be compared: Garros[11] flew from Tunis to Rome at the beginning of 1913, and in September of that year (1913), on a flight from Toulon to Bizerta, he crossed the Mediterranean Sea in an extension of 800 km. On April 24, 1913, Gilbert[12] flew from Paris to Medina (Spain), covering 1,020 km, and a little later Guillaux[13] covered 1,253 km on the Biarritz - Kallum (Holland) route. Not to forget the Paris-Berlin flights of the French pilots (Audemars[14], Daucourt[15], Letort[16]) without a stopover and the epoch-making flight Paris-Berlin-Warsaw of the Frenchman Brindejonc des Moulinais (Morane), which continued via Petersburg, Stockholm, Copenhagen, Hamburg and Hague back to Paris (4,860 km in eight days) and brought stormy honours to French aviation.

The German aircraft industry continued on its chosen path. Thanks to the support of the Nationalflugspende, it had been able to build machines with the highest climb performance, which could handle cross-country and night flights of a total distance of 2,160 km from the starting point without any difficulty, and had also, as already briefly mentioned above, snatched the third important world record: the one for flight duration. It had been newly conquered by Fourny on September 11, 1912 with 13:22:00 for France. On February 3, 1914, it passed into German hands for the first time: Langer[17] flew 14 hours, 7 minutes. That was the prelude, followed in almost bewilderingly quick time by

After National Flugspende

Above: Sablatnigs world record flight with 5 passengers. October 14, 1913: He climbed up to 890 m.

record performances until it reached the record mark on July 11, 1914 with Reinhold Boehm's[18] 24-hour flight. At that time, 20 days before the start of the war, German aircraft and their engines were at the cutting edge of technology.

However, the effectiveness of the Nationalflugspende also extended to the establishment and expansion of flight centres, in particular the support of the "Versuchsanstalt für Luftfahrt" (Aviation Research Institute) and the engine competition for the Kaiser Prize. In all this, the principle was to raise the general level of aviation. In addition, considerable sums of money from the Nationalflugspende were used for aviation competitions, in which the industry, which had almost always lost out in previous competitions due to unnecessary hardships that were not justified by sporting interests, now also had a say instead of benefiting from the further development of what it had won.

Summarizing what was achieved through the Nationalflugspende, it can be said that despite all the initial difficulties, German aviation as a whole and the aviation industry in particular received a decisive impulse.

The plants, which in 1911 were still struggling with their teething troubles, had become stronger and were on a par with, and in many cases even superior to, their foreign competitors. Not all of the early German aircraft companies survived the rapid development. In particular, those companies that did not have the necessary financial foundation a technological background were unable to keeping pace.

A prime example of the development of German aircraft companies is represented by Albatros-Flugzeugwerke. After the company had started with aircraft repairs, and in 1910 had switched to building farm planes, for which it had acquired the license, this machine developed into a completely new, independent type. It went over to the fuselage biplane and progressed from success to success with it.

Most of the German aircraft companies have a similar history; the actual companies existing before the beginning of the war are listed in approximate order of the start of production:

1. Luftfahrzeuggesellschaft GmbH, Johannisthal
2. Rumplerwerke GmbH, Johannisthal
3. Euler, Frankfurt a.M.
4. Albatroswerke, Johannisthal
5. Automobil- und Aviatik-A.G., Mülhausen
6. Otto, München
7. Hansa und Brandenburgisch Flugzeugwerke, Brandenburg
8. Luftverkehrsgesellschaft, Berlin-Johannisthal
9. D.F.W. (Deutsche Flugzeugwerke), Leipzig-Lindenthal
10. Ago-Flugzeugwerke, Johannisthal
11. Halberstädter Flugzeugwerke, Halberstadt
12. Flugzeugbau Friedrichshafen GmbH, Friedrichshafen
13. Kondor-Flugzeugwerke GmbH, Essen

Development of the Aviation Industry During the First World War

Above: Werner Landmann set a new endurance world record with his Albatros biplane with 21 hours 49 minutes and a world distance flight record of 1,900 km.

14. Germania-Flugzeugwerke, Leipzig
15. A.E.G. (Allgemeine Elektrizitätsgesellschaft), Berlin
16. Gothaer Waggonfabrik A.G., Gotha
17. Pfalz Flugzeugwerke, Speyer.

A considerable share of the successes achieved by the German aircraft industry in the competition for the Nationalflugspende prizes was attributable to the engine industry, which in turn had again received considerable support from the Kaiserpreis (endowed in 1912, distributed in 1913). The Emperor's donation of 50,000 M from his private purse was followed by further prize donations: 30,000 M from the Reich Chancellor, 25,000 M from the Minister of War, 10,000 M from the Reich Navy Office and another 10,000 M from the Reich Office of the Interior, so that a total of 125,000 Marks in prizes were offered.

With the increase in military contracts and as a result of funding from the National Flight Donation, a certain upturn in the aviation industry can be observed from 1912 onwards.

Existing companies expanded their production capacities more quickly, also in anticipation of a further increase in military orders due to the tense international situation, and new companies emerged. In view of the large number of new companies that were pinning their business hopes on a future increase in military orders, the Department of Aircraft Industrialists in the Association of German Motor Vehicle Industrialists warned against ill-considered start-ups and asked the Prussian War Ministry for a statement, which was published in May 1912 in the Deutsche Luftfahrer-Zeitschrift:

Statement by the Prussian Ministry of War on May 6:

"In view of the steady increase in the number of aircraft factories, it seems reasonable to fear that only some of the factories will be able to secure a secure existence under the present conditions. For the near future, it must be expected that the army administration will be almost the only customer on the aircraft market. At present it is impossible to say whether the interest in aviation, if it takes hold of wider sections of the population, will lead to the airplane becoming widespread in our sporting life. After the events in France, only limited use will have to be expected in this field for the time being. It therefore seems to the War Ministry to be in the interests of national industry that further aircraft factories should initially only be set up if they are particularly well-funded and generous companies and only those types that are absolutely certain to succeed are built."

Only factories classified as capable of development

After National Flugspende

Right: Euler B.I aircraft in front of Euler works at the Griesheim airfield near Frankfort.

Above: Albatros-Werke, engine installation shop.

Above: Albatros-Werke: Mass production of fuselages.

Development of the Aviation Industry During the First World War

Above: Ago-Flugzeugwerke in Johannisthal.

were supported by the army administration with orders and contracts for training; the aim was to create a number of large, financially strong companies that were capable of mobilization and able to develop the best possible aircraft in free competition with each other using their own resources in accordance with military specifications. The War Ministry encouraged financially strong large corporations such as AEG or the Gothaer Waggonfabrik to diversify into aircraft construction, but without entering into any binding purchase commitments. The General-Inspektion des Militär-Verkehrswesens saw the advantage of such factories in the fact that their strong parent company could keep them alive even without major military orders, but on the other hand production facilities, machines and skilled workers were already available should expansion become necessary.

Nevertheless, the German aircraft industry thus had a free field of work on all directions. And not only that: wide circles were now interested in the industry, which had been so carefully overlooked for so long and was now growing so fast. Not only the actual final aircraft manufacturers developed in leaps and bounds, no, alongside these a whole series of participating suppliers emerged, such as instrument manufacturers, manufacturers of fittings, ropes, seats, wheels, etc. A real jolt went through the entire German industrial landscape.

The procurement policy, the distribution of orders and the delivery conditions had in fact triggered a process of selection and change among the companies: on the one hand, many factories had to give up due to the lack of sales success of their designs and insufficient share capital; on the other hand, sales to the military gave rise to around ten companies that grew into large enterprises with more than 50 employees. The successful army suppliers were able to make a profit, increased their capital, in some cases by founding stock corporations or limited liability companies, and expanded in anticipation of further army orders, even creating overcapacity. In the manufacturing process, the tinkering of the pioneering years was replaced by serial production as production figures rose, and machines were increasingly used for various tasks.

Although the technical perfection of the aircraft was to be the responsibility of the companies, subject to certain quality requirements of the military, the direction and goal of development was to be determined solely by the army and navy authorities.

What the German aircraft industry had gained in terms of

After National Flugspende

Above & Right: From April 1, 1914, Alfred Friedrich was chief pilot at Rumpler-Flugzeugwerke in Berlin-Johannisthal. In June, he caused a sensation with his flights Berlin-Sofia, where he wanted to demonstrate German flying equipment, and Sofia-Bucharest, which was the first overflight of the Balkan Mountains with a passenger. Also noteworthy was his Berlin-Paris-London-Berlin round trip on September 6, 1913, in an Etrich-Rumpler-Taube. The left photo shows Etrich as Friedrichs observer on his flight from Paris to London (the aircraft shown are not identical).

Edmund Rumpler Luftfahrzeugbau GmbH, Production

Year	Production (total)	Deliveries to Army
1911	19 aircraft	10/*11* aircraft
1912	53 aircraft	48/*49* aircraft
1913	63 aircraft	67/*67* aircraft
1914	108 aircraft	NA/*110* aircraft
1914	243 aircraft	*237* aircraft

Source: Figures according to Schwipps, Werner: *Schwerer als Luft*, italicized figures according to *Militärgeschichtliche Fakultät: Die Militärluftfahrt bis zum Beginn des Weltkrieges 1914..*

technical design and manufacture of aircraft through the rich financial contributions from the Nationalflugspende was best demonstrated by the public competitions: the Flugwochen, long-distance flights, etc., of which the flights of Friedrich and Reichelt to Paris have already been mentioned. Of foreign competitions, the water flight on the Upper Italian Lakes saw the German old master Hirth as a contender, who gave the French flying technique a considerable beating by his outstanding performance. Hirth and Ernst Stoeffler had entered the Monaco flight. Both performed outstandingly. The German national competitions (Dreiecksflug, Prinz-Heinrich-Flug, etc.) fostered lively competition among the German companies, which contributed significantly to the improvement of types and the expansion of newly learned skills.

Hand in hand with this, of course, went an expansion of the existing aircraft factories, which switched from the mostly manual work hitherto used to series production. The much more subtle and cleanly executed work has always been an advantage of German aircraft over foreign ones. We need only recall the triumph of the Albatros and Aviatik

Development of the Aviation Industry During the First World War

Above: A forerunner of the Aviatik P.13 (1913/14), which was later built in series.

companies at the Paris "Salon" in 1911.

The main customer for the aircraft companies was, of course, the army administration. At the beginning of the war, the acceptance requirements were still the same as in 1913. According to these requirements, the aircraft, powered by a 100 hp engine, had to be able to carry fuel for four hours and a payload of 200 kg, including two passengers, and at the same time develop a speed of at least 90 km/h and a climb rate of 1,000 m in 10 to 15 minutes, while limiting the take-off distance to a maximum of 100 m and the take-off run to a maximum of 70 m. The aircraft had to be able to fly at a maximum speed of 90 km/h and a climb rate of 1,000 m in 10 to 15 minutes.

For the growth, if not the existence of a company, it was crucial to be able to win orders from the military. An example of two manufacturers who offered a similar type of aircraft will illustrate this: The Etrich-Rumpler Taube, which became the best-known aircraft of its time in Germany, was able to achieve great success from 1910 onwards - after Igo Etrich's patent protection was revoked, 23 companies built Tauben with the characteristic wing shape. The city inspector of Hanover, Karl Jatho, who had built several flying machines from 1903 onwards, also designed a Taube, with which he won the prize for the first circuit of Hanover in the air in 1912. The following year, he founded Hannoversche Flugzeugwerke GmbH, whose commercial basis was to be the production of the Jatho XI Stahltaube for the military, a model that was equipped with a 100 hp engine and military controls in accordance with military requirements. However, after the army administration did not order this model, the factory soon ran into financial difficulties and had to be closed down again in 1914.

Edmund Rumpler Luftfahrzeugbau GmbH, on the other hand, developed into one of the largest airframe manufacturers in Germany before the First World War thanks to army orders for various models of its Taube. The company's economic basis was the Taube and its further developments; a successful military biplane only appeared shortly before the start of the war.

Although Rumpler-Werke also received a few orders for its Taube from the civilian sector, the majority of orders came from the army administration:

Rumpler became the largest supplier of monoplanes for the army, and the company expanded constantly in anticipation of further military orders, as can be seen from the number of people employed in the workshops: in 1909, 18 (*8*) workers were employed, in 1910 already 40 (*50*), in 1913 the number was 245 (*245*) and in 1914 then 630 (*400*). Although the figures on annual army deliveries and the number of employees differ in the literature, they show the same trend that also applies to the other companies used by the army as suppliers: the vast majority of the aircraft produced went to the military. Günter Schmitt speaks of around 90%; the number of aircraft built and the number

Size of Operational Air Forces in 1914			
Country	Aircraft	Airships	Captive Balloons
Russia	263	4	46
Germany	232	4	16
France	165	10	10
Great Britain	63	–	–
Austria-Hungary	48	3	6
Belgium	16	2	2
Source: Groehler *"Geschichte des Luftkriegs"*			

of employees rose sharply within four years, especially from 1912 onwards. A lack of orders from the army, on the other hand, often meant the end of a company, or at least prevented greater growth.

When the First World War broke out at the end of July, there was still no air force and he initially joined the Imperial Air Force as a contract employee; only later did he receive a military rank.

Further examples of this development can be cited: Without orders from the army, the Dorner-Flugzeug-Gesellschaft had to give up after a short existence in mid-1912, and the Harlan-Flugzeugwerke and the Bussard-Werke suffered a similar fate. Smaller companies like: Erste Magdeburger Fliegerwerke und Fluggesellschaft, Magdeburg, Etrich-Fliegerwerke GmbH, Liebau i. Schlesien or Nordwestdeutsche Flugzeugwerke Evers & Co., GmbH, Bremervörde b. Bremen could not survive a long to their powerful rivals.

Despite diligent efforts, Hans Grade Fliegerwerke, founded in 1909, was also unable to secure any orders from the military.

The company, which focused on light aircraft designs, existed until 1917 and produced around 80 civilian aircraft in the pre-war period, but these were only small numbers compared to the production of the large companies involved in army deliveries. Albatroswerke, on the other hand, flourished and developed into one of the largest manufacturers in pre-war Germany, as it was one of the most important suppliers to the army. Albatros was also successful in the civilian sector, but the majority of the aircraft were produced for the military: in 1912, 46 (*48*) of the 55 aircraft built went to the Prussian air force, and in 1913, 85 (*89*) of 96 aircraft, plus five monoplanes, went to the Bavarian air force. In the following year, 336 (*338*) aircraft were produced. The factory facilities were expanded several times and the number of employees rose from 270 (*250*) in 1912 to 650 (*745*) in August 1914. The companies that were among the suppliers to the army before the outbreak of war and which were helped to grow by military orders also included AEG, Automobil und Aviatik AG, Deutsche Flugzeug-Werke GmbH (DFW), Flugmaschinenwerke August Euler, Fokker-Aeroplanbau GmbH, the aircraft works department of Gothaer Waggonfabrik AG, Emil Jeannin Flugzeugbau GmbH, Luft-Fahrzeug-Gesellschaft mbH (LFG) and Luft-Verkehrs-Gesellschaft AG (LVG).

Germany entered the First World War with these requirements, which were extremely low in comparison with our aircraft industry, which was already very highly developed at the time.

Under the impression of the second Moroccan crisis and the Balkan wars, which carried the latent danger of an expansion to other European powers due to various alliance obligations, as well as the rapid progress of the armament of France and Russia, Germany also planned a new rearmament for 1913 since the foundation of the Reich. The air force was greatly expanded and reorganized in terms of personnel and material within the financial framework established in negotiations with the Reich Treasury and with the support of funds from the National Flight Donation: On October 1, 1913, the Royal Prussian Air Corps, which had emerged a year earlier from the Training and Experimental Institute for Military Aviation, was dissolved; in its place, four Prussian air battalions were established, each assigned to an army corps.

The aviation and airship units were subordinate to the Inspectorate of Aviation Troops (Idflieg), also created on October 1, 1913, and the Inspectorate of Airship Troops, respectively, both of which came under the command of a technical department, the Inspectorate of Military Aviation and Motor Vehicles, which in turn remained subordinate to the General Inspectorate of Military Transport. This assignment did not correspond to the self-image of the Inspectorate of Aviation Troops.

In the German Reich, the first companies to manufacture flying machines for profit emerged from 1908 with the factories of the pioneering industrialists August Euler, Hans

Endnotes
1 Arthur Adolf Graf von Posadowsky-Wehner Freiherr von Postelwitz (* June 3, 1845 in Groß-Glogau; † October 23, 1932 in Naumburg (Saale)) was a German politician.
2 Theodor Lewald (* August 18, 1860 in Berlin; † April 15, 1947 ibid) was a high-ranking administrator of the German Reich, spokesman for the Reich government in the German Reichstag, member of the Executive Committee of the International Olympic Committee, German sports official and chairman of the Organizing Committee of the 1936 Olympic Games.
3 Alfred Hildebrandt (* June 10, 1870 in Wittingen; † February 24, 1949 in Oberkochen) was a German aviation pioneer and writer.
4 Marcel-Georges Brindejonc des Moulinais (* February 18, 1892 - † August 5, 1916) was a French pilot best known for several long-distance flights, such as the

overflight of the Baltic Sea.
5 Viktor Stoeffler (* June 9, 1887 - † July 1, 1947), license number 174, was a pioneer of French aviation. He was a test pilot at Aviatik in Mulhouse, Freibourg (1912-1913) and its later factorydirector, then in Leipzig. (1914-1918).
6 Ernst German Schlegel (* June 21, 1882 in Konstanz; † January 4, 1976 ibid.) was a German aviation pioneer, aircraft designer and engineer. In 1909, he moved to Mainz to attend the engineering school there. Based on his own designs and plans, he constructed the "Schlegel-Züst flying machine"together with his Swiss friend Robert Züst. Schlegel was a mail pilot, test pilot, water pilot, and then a wartime pilot in World War I with numerous missions. He was promoted to lieutenant and received the Iron Cross 1st and 2nd class. Later he was a senior technical employee at Rumpler-Flugzeugwerke in Berlin and in 1917 was transferred by the army administration to the Pfalz aircraft factory in Speyer aschief engineer.
7 Karl Christian Maximilian Caspar (* August 4, 1883 in Netra (Hesse-Nassau); † June 2, 1954 in Frankfurt-Höchst) was a German pilot, aircraft manufacturer and lawyer. He gained notoriety, especially in the 1920s, for the types of aircraft developed and built by Caspar-Werke. As an airplane pilot who took his exam before the First World War, he is counted among the Old Eagles.
8 Robert Thelen (* March 23, 1884 in Nuremberg; † February 23, 1968 in Berlin-Hirschgarten) was a German engineer, pilot and aviation pioneer. During World War I, Thelen was employed by Albatros Flugzeugwerke as a test pilot and designer; at the same time, Ernst Heinkel was responsible for designing bomber aircraft. Among Thelen's designs was the Albatros D.III, one of the Air Force's most widely used fighter aircraft. As a test pilot, he flew the Albatros seaplanes built at Friedrichshagen.
9 L'Auto, 2.000 Kilometres en 24 heures, Mercredi 15 Octobre 1913.
10 The *Aero*, published in London.
11 Eugène Adrien Roland Georges Garros (* October 6, 1888 - † October 5, 1918) was a French aviation pioneer and successful fighter pilot during World War 1.
12 Sub-Lieutenant Eugene Gilbert (* July 19, 1889 - † May 17, 1918) was a World War I flying ace (five aerial victories). He was also a famous pioneer of pre-war racing and flew in many countries in Europe.
13 Ernest François Guillaux (* January 24, 1883 - † May 21, 1917), better known by his adopted name Maurice Guillaux, was a French airman who, among other things, spent 7 months in Australia in 1914, where he carried the first airmail (Melbourne - Sydney) between July 16 and 18, 1914.
14 Edmond Audemars (* December 3, 1882 in Le Brassus - † August 4, 1970 in Deauville) was a cyclist, aviator and Swiss industrialist.
15 French aviator Pierre Daucourt won, among others, the Pommery Cup for the longest direct flight between sunrise and sunset. He covered an estimated distance of about 570 miles. Daucourt took off at 5:59 a.m. from Valenciennes, near the Belgian border, and flew directly to Biarritz, near the southwestern end of France. He arrived there at 5:38 p.m. He made three stops to refuel his plane.
16 Émile Louis Letord, sometimes Letort (* March 10, 1880 - † March 28, 1971) was a French industrialist and aviation pioneer. He merged the companies Letord and Niepce, thereby creating the aircraft company Société d'Aviation Letord.
17 Bruno Langer (* May 2, 1893 in Bützow/Mecklenburg - † October 19, 1914) received aircraft pilot certificate No. 203 on Rumpler-Taube (October 19, 1912).
18 Reinhold Boehm (* July 7, 1890 in Gerlauken/East Prussia) did not receive his pilot's license until April 18, 1913.

During the War

The Development of German Aircraft Industry During the War

Grade and Edmund Rumpler, initially often licensed or unlicensed replicas of proven foreign designs.

In addition to the commercially oriented companies that manufactured for sale and whose founders and owners were often not aviators themselves, individual designers and aviation enthusiasts developed and built flying machines of their own design with their own resources in the pioneering period between 1909 and 1914, most of which remained one-offs. The representatives of "heavier than air" flight technology began to emancipate themselves from the airship industry, which also manifested itself in the establishment of flight technology sections in the German Airship Association (Deutscher Luftschiffer-Verband) and finally the founding of the independent Association of German Flight Technicians (Vereins Deutscher Flugtechniker) in 1909. Until the outbreak of war, the number of aircraft companies rose sharply, especially in the period after 1912, when entrepreneurs hoped for better chances of winning lucrative military contracts in view of the plans to expand the air force. The financially strong large-scale industry now also showed more interest in aircraft construction, with Siemens-Schuckert-Werke, Allgemeine Elektrizität-Gesellschaft (AEG) and Gothaer Waggonfabrik AG being the first large companies from outside the industry to set up aviation technology departments. A number of large companies and bankers, AEG, Borsig, Krupp and Leo Delbrück, had already contributed financially to the foundation of Flugmaschine Wright GmbH initiated by the Motorluftschiffahrt-Studiengesellschaft in 1909, but most of the early company founders came from the middle class, often from technical professions. By mid-1914, the number of aircraft companies and workshops amounted to over 50, but only a small number of them remained in existence for any length of time.

As already mentioned, the military authorities showed an early interest in aircraft as a new means of warfare, and the militarization of aviation began. Not least due to the pressure of the arms race, land and naval forces tested balloons, airships and aircraft, built up an organizational structure and acquired aircraft, mainly from private industry. The aircraft industry, which could only tap into a narrow market for its product in the civilian sector that allowed little or no expansion, saw the armed forces as the institution that seemed to have a use for aircraft in large numbers,

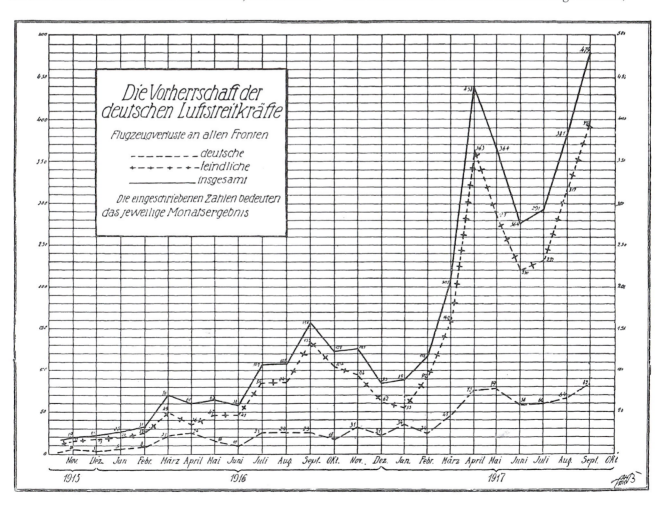

Above: Number of aircraft losses in accordance to German description ("Luftwaffe 1918).

Development of the Aviation Industry During the First World War

Above: A wingless Albatros C.I.

the most important customer, a promoter and financier of expansion, and thus willingly turned to the military. Due to the small non-state sales market, investment in the expansion of the industry had to come primarily from the public sector and took the form of orders for military aircraft, but not direct subsidies. The tense global political situation in the last few years before the outbreak of war and the intensification of armament efforts by the major powers led to an expansion of the aviation industry. In anticipation of lucrative military orders, existing plants expanded, new companies were founded, and companies from outside the industry and large corporations began to diversify into the now increasingly profitable business of air armament by setting up aircraft construction departments. The industry became dependent on the military: its position as by far the most important consumer, holding an almost monopoly position, enabled the military administrations to exert extensive influence on the structural development of the industry and the technical development direction of the aircraft - only the military placed orders on a scale that provided the companies with an economic basis and also promised expansion with the aim of increasing profits.

Through specifications and competitions, the procurement authorities steered the technical development of the aircraft in a direction that met the requirements as a means of war, and through the policy of awarding contracts, the structural development of the industry. As the most important consumer and paymaster, the military was able to force private companies to adapt to its requirements and largely determined the size, productivity and profitability of the industry. The high degree of orientation towards the war ministries as customers made the aviation industry less dependent on the free economic market than on a politically determined market, determined by the prevailing military doctrine, the global political situation and the assessment of the national security situation.

The British Army relied primarily on private aircraft companies and companies from outside the industry to replicate designs from the state-owned Royal Aircraft Factory, while the Royal Navy relied on aircraft developed and manufactured by private companies.

After unsuccessful attempts at state aircraft construction, however, the German air forces obtained their material exclusively from private factories, and the military authorities endeavored to promote the development of large, efficient aircraft factories through their procurement policy.

In the First World War, aircraft alone were not the decisive weapon in the war, the decision was made in the huge material battles on the ground; the purpose of the majority of the air forces was ultimately to provide direct or indirect support for the ground troops on the battlefield. For the airborne troops, the "Great War" was primarily an "army

Above: Aeg C.IV (C.1044/16) being inspected by army commanders.

cooperation war"; as force multipliers, they supported their own army in the fight against enemy armies by conducting aerial reconnaissance and directing artillery fire, by protecting their own reconnaissance aircraft, airships and balloons and by interdicting enemy reconnaissance by armed aircraft, by attacking ground targets on the battlefield and behind the front line with airborne and airdrop weapons and by repelling air attacks.

In the course of the First World War, the airplane became increasingly important in its function of supporting the ground war. With the expansion of war operations, but also as a result of high losses due to accidents and enemy action, the demand for military aircraft increased massively. In an effort to achieve the ambitious production targets set by military planning, or at least come close to them, the aviation industry greatly expanded its design and manufacturing capacities up to the end of the war. Existing aircraft factories expanded, new companies were founded and, in Great Britain in particular, manufacturing capacity was increased through the increased involvement of non-industry copying companies, which participated in the armaments business by temporarily diversifying into aircraft construction.

The production processes began to change and with the high quantities, series production using machines became more widespread. In the course of the mobilization of the entire economy, which was increasingly considered necessary for the material-intensive war, the state further expanded its influence on the aviation industry through supportive and guiding measures. In the shortage situations that arose, the state ensured the provision of labor and raw materials for aircraft construction, which was given high priority in the armaments industry, and provided financial support through payment arrangements. In the interests of safeguarding the flow of production, state bodies began to intervene in internal industrial affairs, which had previously been regarded as sacrosanct. Intervene. Beyond questions of technology and quality assurance of the aircraft product, they also intervened in the employer-employee relationship, the management of raw materials and entrepreneurial independence. During the demobilization that began after the armistice in November 1918, the armed forces were quickly reduced to a level that was determined by the political side based on the changed military requirements of peacetime.

Aircraft enjoyed a special position in the organization of material procurement: at the request of Lieth-Thomsen, Ludendorff, First Quartermaster General in the 3rd OHL

Development of the Aviation Industry During the First World War

Above & Below: Hansa-Brandenburg CC flying boat. The aircraft was finally equipped with two guns. The Benz engine got its own cowling to reduce drag and prevent damage by salt water.

and a long-time supporter of aviation, separated them from the area of responsibility of the Weapons and Ammunition Procurement Office (Wumba) and transferred responsibility to the Commanding General of the Luftwaffe (Kogenluft) and Idflieg, which ensured more direct control and influence over material procurement.

The Idflieg's Flugzeugmeisterei was responsible for the procurement, acceptance and dispatch of aircraft, while special technical departments such as the Weapons Command or the Radio Telegraphy Command were responsible for questions of armament and equipment. In August 1918, the Kogenluft ordered a streamlining of the

Above: A Friedrichshafen FF 19 is placed on wheels to be rolled out of the construction hall.

procurement apparatus in order to ensure the best possible utilization of the remaining resources:

"The Commanding General of the Air Force has decreed that the current Flugzeugmeisterei will be responsible for the entire procurement and supply of all material to the front in future.

Among others, the Flugzeugmeisterei will include: Bombs Department, Weapons Department, F.T. Department, Raw Materials Department, Air Transport Department, Weapons Service Exemption Department, Central Investigation Commission, Patent Department, the Air Force Commissioner. (German meaning accordingly: Kdo. Bomben, Kdo. Waffen, Kdo. F.T., Rohstoffabteilung, Fliegertransport-Abteilung, Abteilung Waffendienstbefreiung, Zentraluntersuchungskommission, Patentabteilung, der Beauftragte der Luftstreitkräfte).

In Germany, the competition between the army and navy in aircraft procurement was not as pronounced as in Great Britain, with the Prussian authorities dominating the procurement system. The naval aviators, who largely concentrated on the war at sea, had to cover their limited requirements for land-based aircraft via the relevant army departments; the Reichsmarineamt procured seaplanes mainly from a small number of specialized companies, mainly Hansa- und Brandenburgische Flugzeugwerke AG and Flugzeugbau Friedrichshafen GmbH, which alone supplied around 40% of all seaplanes.

Measured against the steady state in the summer of 1914, the navy's air forces experienced a similar expansion to that of the army's air forces. At the beginning of the war, naval aviation was still in the experimental stage: the naval aviation department had a small number of seaplane stations on the North and Baltic Seas and around 30 land and seaplanes, mainly unarmed single-engine biplanes with floats and a crew of two, although only around half of these aircraft were considered conditionally operational. In cooperation with companies specializing in the construction of seaplanes, the Seaplane Experimental Command pushed ahead with the technical development, and the seaworthiness, reliability, flight performance and carrying capacity were rapidly and significantly increased. Initially, the main task of the seaplanes was coastal surveillance against surface ships and submarines and observation flights to protect the ships entrusted with clearing mines in the waters off the German

Development of the Aviation Industry During the First World War

Above: Kaiserliche Werft Kiel Marine-Nummer 463 (Type 462) with Benz engine.

naval bases from surprise attacks, but new two-seater seaplanes with a longer range and a higher payload, which allowed radios and weapons to be carried, were also used for reconnaissance up to the south coast of Great Britain and for bombing and machine gun attacks on enemy ships. The use of torpedoes was tested, but despite some success, the results were not convincing and the trials were ended in 1918.

By November 1918, the navy's air force had grown to around 30 land-based air formations, with over 30 naval air stations on the North Sea, Baltic Sea, Black Sea and Mediterranean. During the war, a total of 2,138 seaplanes were delivered to the naval air forces and the number of personnel increased from 217 to around 16,000 men. At the end of the war, the army's air forces had 2,709 operational front-line aircraft.

In March 1918, the strength on the Western Front had still amounted to over 3,600 aircraft, the air force was at its quantitatively highest steady state, but the German spring offensive and the Allied counterattack had claimed massive losses, which the aviation industry was unable to compensate for despite the increased output.

As the fronts became increasingly entrenched during the First World War and it finally came to positional warfare, reconnaissance over the battlefield and the enemy's nearby hinterland played an increasingly important role. This was the only way to recognize the enemy's intentions in good time. For this reason, the army increasingly demanded timely reconnaissance results from the air force. Specialized aircraft were needed to successfully perform these tasks, and the C-types were created, single-engine, armed biplanes, usually with a camera for aerial photography and radio equipment. The crew consisted of the pilot and the observer, who had to operate the camera and operate the machine gun to defend against enemy aircraft when they appeared. The German aircraft manufacturers threw themselves into this task with the result that a myriad of different models were developed. These included the Ago C.I, the Albatros C.I, the FG Roland C.II, the Otto C.I, the Rumpler C.I, the AEG C.IV, the DFW C.V and the LVG C.V, to name but a few. In addition, there was a considerable number of prototypes that had been developed without a contract in the hope of getting a slice of the big armaments pie. By 1916, the variety of types had become so great that there were serious problems with maintenance and the provision of spare parts. The leadership of the air force was forced to eliminate the confusion of types and rely on the best models, such as the Albatros C.V and DFW C.V. These models had been available since 1916 and 1916 respectively. These models had been in series production since 1916 and 1917 respectively and met the requirements of the Fliegertruppe. Only the top speed, which was around 170 km/h, caused increasing problems with the emergence of faster enemy fighter monoplanes, so that in the summer of 1917 the demand for a faster C-aircraft was made to the industry.

Above: Crews of Rol C.IIa "Walfisch" and LVG C.IV with Daimler engine (left).

Number of Aircraft Built in Germany									
Group	1911	1912	1913	1914	1915	1916	1917	1918	Total
A	11	60	168	294	13	22	–	–	568
B	13	76	278	1,054	1,312	440	2,993	25	6,191
C					2,674	2,726	10,337	7,320	25,057
D					1	2,129	4,945	5,132	12,207
Dr							338	1	339
E					347	300	–	381	1,028
G, R					185	465	589	789	2,028
J							450	463	913
N						100	94	10	204
S								2	2
Total	24	136	446	1,348	4,532	8,128	19,746	14,123	48,537

Development of the Aviation Industry During the First World War

3.5 The Influence of the State and Its Military Institutions on the Development of the Aircraft Industry

The influence of the German state on the development of the aviation industry was extremely hesitant. People were proud of their Zeppelins and initially left aviation to sportsmen, adventurers and other enthusiasts.

Thanks to the associations for the promotion of aviation in general, which were spread throughout Germany, the West German associations in particular supported the efforts of aircraft manufacturers to catch up with the aviators of neighboring France.

In 1911 (20th to 28th May), the "First German Reliability Flight on the Upper Rhine" was organized and carried out with the aim of first determining the quality of German aircraft. As early as 6 August 1910, when the southwest German airship clubs formed a cartel at the invitation of the Frankfurt Airship Club, the plan emerged to use the favorable location of the clubs in the Rhine-Main plain, which was virtually predestined for aviation purposes, for an air show. The military representatives were prominently represented, including taking the chairmanship of the event.

It was of decisive importance for the realization of the first German Reliability Flight on the Upper Rhine when His Royal Highness Prince Henry of Prussia, who had made contact with aviation circles the previous autumn through his flight exercises under the direction of August Euler, declared his willingness to take over the protectorate of the flight in February. As the brother of Emperor Wilhelm II, his active participation in the organization of the competition was all the more valuable. The prince, a constant supporter in the following years, supported the organizers on the working committee and chaired important meetings with aircraft manufacturers himself.

In the 10 months available to prepare the event, a myriad of difficulties had to be overcome. In practice, the state authorities at various levels proved willing to make the project a success.

No technical requirements were placed on the aircraft themselves. An acceptance committee decided on admission to the competition after a preliminary flight. Eight pilots competed against each other to cover the 650 kilometers within 10 stages.

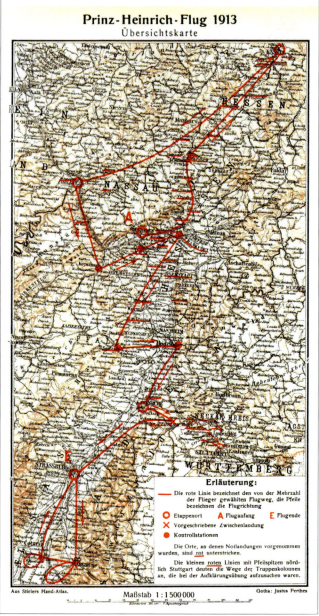

1. 151,350 marks were raised for German aviation.
2. 14 cities in the Upper Rhine Plain were offered flight demonstrations, thereby promoting interest in aviation in wide circles.
3. Seven aviation clubs had the opportunity to participate in a major air sports event, thereby enhancing their reputation in sporting circles.
4. For the first time, an attempt was made in the competition for a cross-country flight to reward not the sporting performance but primarily the reliability of the aircraft by requiring that the entire flight distance be covered by the same aircraft.
5. it proved to be a good idea to consider the total flight duration of the individual stages, i.e. including all intermediate landings, repairs, etc., as a measure of reliability.
6. for the first time, German officers took part in a public competition as pilots.
7. a special weather service for cross-country flights was introduced for the first time.
8. it turned out that monoplanes with elastic wings were superior to rigidly built biplanes of the previous design in turbulent air.
9. the aircraft engines manufactured in German and Austrian factories proved to be equal to the French rotary engine in

Influence of the State

```
         A n f o r d e r u n g e n
         ---------------------------
         an Kriegsflugzeuge für 1913.
   (übersandt von der Inspektion des Militär-Luft-u.Kraftfahrwesens).

 1. Deutsches Material und Fabrikat in allen Teilen.
 2. Guter Sitz für Führer und Beobachter, leichte Verständigung zwi-
    schen beiden; Steuerorgane für den Führer.
 3. Für die Besatzung wird möglichst großer Windschutz, bequemer Sitz
    und völlige Armfreiheit verlangt. Die Karosserie muß genügend Raum
    zum Einbau einer Abwurfvorrichtung und Unterbringung von Abwurfbom-
    ben, sowie zum unbehinderten Photographieren besitzen.
 4. Möglichst automatische Stabilität und mühelose Betätigung der
    Steuerorgane.
 5. Abweichungen von der Militärsteuerung bedürfen besonderer Abma-
    chung.
 6. Uebersichtliche Anordnung der Instrumente (Barometer, Barograph,
    Kompaß, Tourenzähler, Stoppuhr). Prüfungsmöglichkeit für den Benzin-
    und Oelstand durch den Führer im Fluge muß vorhanden sein.
 7. Eigengeschwindigkeit von mindestens 90 km. Bei dieser oder größeren
    Geschwindigkeit muß ihre Herabsetzung während der Fluges bis auf
    75 km möglich sein, ohne die Flugfähigkeit zu beeinträchtigen, d.h.
    in wagerechter Lage geradeaus fliegen zu können.
 8. Größte Breite :   14,5
    Größte Länge  :   12,-
    Größte Höhe   ;    3,50
    mit Rücksicht auf die Unterbringung.
 9. Betriebsstoffe für 4 Stunden.
10. Motorstärken nicht über 100 PS. Abweichungen unterliegen der Geneh-
    migung der Heeresverwaltung.
         Bei gleichwertigen Leistungen werden Flugzeuge mit schwäche-
    ren Motoren bevorzugt.
11. Sichere und gefahrlose Unterbringung der Betriebsstoffe (Betriebs-
    stoffbehälter über oder hinter der Besatzung sind ausgeschlossen.)
12. Anlaßvorrichtung bezw. Andrehvorrichtung.
13. Spielraum für Propellerspitze nicht unter 45 cm. vom Boden.

                                - 2 -

14. Steigfähigkeit mindestens 800 m in 15 Minuten.
15. Anlauf bis höchstens 100 m auf ebenem Boden. (Startmannschaften ge-
    stattet). Auslauf höchstens 70 m; Wendigkeit auf dem Boden.
16. Abflugsmöglichkeit und Landen auf dem militärischen Probefeld in
    Döberitz (bei Typenabnahme).
17. Nutzlast (außer Betriebsstoffen, Instrumenten und Werkzeug) von
    mindestens 200 kg. (Führer und Beobachter sind hierin enthalten).
18. Gleitflug aus 500 m Höhe (mit Rechts- und Linkskurven) mit abge-
    stellter Zündung. (Bei Typenabnahme).
19. Schnelles Zusammensetzen und Zerlegen, als Norm gilt mit 5 Mann
    Montage 2 Stunden, Demontage 1 Stunde, leichte Verladefähigkeit auf
    Eisenbahnwagen und Landfahrzeugen. Profilfreiheit für Eisenbahn- und
    Straßentransport.
20. Unempfindlichkeit gegen Witterungseinflüsse.
21. Leichte Auswechselbarkeit einzelner Teile (z.B. Fahrgestell).
22. Eine Einrichtung zur vorübergehenden Dämpfung des Motorgeräusches.

                         ----------------
```

Above: The design regulations for flying machines of the German Army of 1913 consisted of 22 requirements.

1914 — Ein Eindecker der Luft-Verkehrs-Gesellschaft (10 Mann Probebelastung).
A monoplane of the Luft-Verkehrs-Gesellschaft (Air Traffic Company) (10 men).

1918 — Ein verspannungsloses Fokker-Jagd-Flugzeug D VIII (24 Mann auf der Tragfläche).
A braceless Fokker one-seater D VIII (24 men on carrying surface).

1922 — Ein Junkers-Flügel in der Forschungsanstalt Dessau: 42 Mann auf einem Flügel.
A Junkers plane in the Research Department Dessau: 42 men on one wing.

terms of reliability.

It is important to emphasize the fact that everything that was achieved was raised by the industrialists themselves, apart from the funds to secure the competition.

The first German reliability flight on the Upper Rhine in 1911, also the first German cross-country flight on a large scale, had set itself the task of testing aircraft and pilots over longer distances in the field and encouraging German industry and pilots by offering high cash prizes.

In the opinion of the organizers and the experts, this purpose was definitely achieved; the lead that France undoubtedly had in the field of aviation could of course not be caught up with, but it could be reduced.

Although not all wishes and expectations could be fulfilled, it was intended to repeat such a competition under the direction of the Southern Group of the German Aviation Association in the following year.

Interest in this competition also grew on the military side. The active participation of the German state continued to be limited, despite the knowledge of the technological lead of the French aircraft industry. In practice, however, it became clear that there were no structural requirements for a military aircraft.

After the Kaisermanöver in September 1912, in which 14 airplanes were initially registered, it had become clear that in an emergency the losses due to material damage could not be fully compensated. As a result, the strength of the Flieger detachments was increased from six to eight flying machines, and the acquisition of reserve aircraft and engines was also called for. The practical results of the evaluation were that the biplane design was superior to the monoplane in terms of takeoff and landing characteristics. With regard to the field of view for the observer, it was found that the visibility in

Development of the Aviation Industry During the First World War

Above: DFW C.V C.9036/16 reconnaissance aircraft.

the monoplane could only satisfy operational requirements. For tactical observation tasks, the biplane design was better suited, especially with a pressurized propeller. The front-seated observer had an excellent, almost unrestricted, field of view and, after the aircraft were later equipped with swiveling machine guns, a correspondingly large field of fire. Reference was also made to the need for rapid disassembly and assembly of the aircraft in the field. Since Otto had not succeeded in further developing his designs and supplying more powerful machines in the course of the year, the Bavarian Ministry finally opened up to competition. Not least because of the fluctuating manufacturing quality. Standardized testing of flying machines did not take place at all until 1913. It was thanks to von Brug that he introduced the BLV (Bau-und Liefervorschriften), or Design and Delivery Regulations. Now the flying machines had to be tested in four load cases before any approval was granted.

With the actual course of the war, a war lasting more than four years on two main and several secondary fronts, and the increasing importance that the German commanders also attached to air support for the ground troops, especially in positional warfare, the demand from the army and navy for aircraft of various categories rose sharply. In addition, the German Empire's allies requested support in the form of material supplies: Turkey received around 400 aircraft from German production during the war, and Austria-Hungary also endeavored to acquire aircraft in Germany, as the production capacity of the Dual Monarchy could not be expanded to meet the steeply rising demands.

On August I, 1914, Idflieg telegraphed all aircraft factories used for army deliveries to increase production to the maximum and to make preparations for an expansion. Although the industry was able to multiply its production capacities through various measures, the high quantitative requirements of the military planners could only rarely be met. The military leadership demanded that the aviation industry increase its output of aircraft and engines with all its might, and the state generally took measures to create a favorable environment for increasing the production of the armaments industry and to increase the efficiency of the companies. As the war progressed, the view prevailed among the decision-makers in the state that all available production capacities had to be geared towards war requirements and that, without deeper intervention by state authorities, this war with its high material consumption could not be continued for much longer, let alone won.

The decision-makers were increasingly the military, the 3rd OHL under Field Marshal Paul von Hindenburg and Lieutenant General Ludendorff, appointed in August 1916, had far-reaching powers, a de facto development towards a military dictatorship began, the civilian administration capitulated to the military. Hindenburg and Ludendorff, the almost legendary victorious commanders of the battles on the Eastern Front, the "Dioscuri", who embodied the hope of ending the war through a military victory and not just through a negotiated peace, held the view that the military leadership must also have the power to decide on questions of domestic and foreign policy if this total war was to end victoriously. In order to achieve a victorious peace, all of the nation's reserves of strength would have to be

Above: DFW R.II with 4 x 260 hp Daimler D.IVa engines (1918).

mobilized due to the material strength of the enemy. Based on the experience gained in the huge material battles of Verdun and the Somme, the OHL formulated an expansion program in August 1916, the Hindenburg Program, which called for the production of weapons and ammunition to be doubled or quadrupled in order to counter the opponents' material expenditure. The air force leadership wanted to increase the number of front-line aircraft from around 1,500 in the fall of 1916 to over 2,300 by the spring of 1917 by increasing the proportion of D and G aircraft, for which an increase in monthly production to around 1,000 aircraft and 1,500 engines was considered necessary. In order to implement its expansion plan, the OHL sought to enforce the primacy of the armaments industry. Control measures were to enable the industry to fulfill the large number of large orders and to considerably increase the production of armaments; human and material resources were to be used with the greatest possible efficiency due to the shortages that arose: Companies producing armaments in bulk were rigorously given preference in the supply of materials and personnel; in December 1916, the Standing Committee for the Consolidation of Factories was set up to shut down companies that were not essential to the war effort and small companies that were operating inefficiently, thus providing the large armaments factories with more materials and manpower.

The law on national service, which was passed in December 1916 following socio-political concessions to the trade unions, was intended to combat the problem of the shortage of manpower and increase the number of workers available for the war effort through redistribution. This law stipulated that the entire male population was to be enlisted for the war effort through a general obligation to work, but was unable to meet the high expectations with regard to easing the personnel situation. The war economy management apparatus was reorganized and institutionally concentrated in order to make it more manageable and efficient: the War Office was created, headed by General Wilhelm Groener, to which the authorities responsible for the war economy, such as the War Raw Materials Department, the War Replacement and Labour Department, which was responsible for the conscription and release of workers, and the Weapons and Ammunition Procurement Office, were assigned. The lack of authority and friction with both the Prussian War Ministry and the military command authorities hampered the work of the War Office as the leading decision-making center, but the measures implemented contributed to a considerable increase in

Development of the Aviation Industry During the First World War

Above: Repair shop at AEG Flugzeugwerk (1917). On the left side are several AEG C aircraft under repair.

armaments production in accordance with the demands of the Hindenburg Program. The price for this, however, was an extreme exploitation of the human and material resources of the German Reich, and in the long term there was a threat of overstretching. (Refer to Morrow, John H.: *German Airpower in World War I*.)

The central role in the move away from market economy principles and the development towards planned economy structures was played by the shortage of raw materials essential to the war effort as a result of the restrictions on foreign trade and the Allied naval and trade blockade imposed with Britain's declaration of war on August 4, 1914 - in the last years before the war, Germany had imported over 40% of its industrial raw materials, and a survey of 900 large industrial companies in August 1914 had shown that the available stocks would only last for half a year on average. Due to peacetime politics, there was no authority in the military bureaucracy at the beginning of the war that took care of the supply of raw materials to industry and their distribution in accordance with the importance of the companies to the war economy. Suggestions for compulsory management soon came from outside the government and the military, from industry: Walther Rathenau and engineer Wichard von Moellendorff from the AEG conglomerate demanded the centrally controlled management of certain raw materials from Prussian War Minister Erich von Falkenhayn as early as the beginning of August 1914. The proposals were quickly implemented with the establishment of the War Raw Materials Department (Kriegsrohstoffabteilung - KRA) in the War Ministry on 13 August 1914, of which Rathenau was appointed head and whose primary task was the organized management of shortages of raw materials essential to the war effort, the control of the development and production of synthetic substitutes and the exploitation of the economic resources of the occupied territories. A system of priorities was established; according to the new distribution regulations, only those companies that were useful to the war effort were entitled to scarce raw materials - a significant step towards mobilizing the economy for total war. As sub-organizations for "absorbing, storing and distributing" the raw materials, the Kriegsrohstoffgesellschaften (War Raw Materials Companies) were formed by merging assigned industrial sectors, the number of which rose to around 200 by the end of the war. The Kriegsrohstoffgesellschaften were largely autonomous organs of self-administration in the industry, private companies which, as trustees and controllers of the state in the organization of raw materials management, were endowed with state executive power, but each also included a state representative who represented the interests of the state and had a right of veto. For the state, which did not have an appropriately competent bureaucracy, the self-organization of the industry under the supervision of the state represented a thoroughly pragmatic solution to the question of organizing raw materials management, even if there was a risk that big industry would exploit its powers for silent predatory competition, price fixing and syndication.

The coordinating and steering measures of the authorities proved successful, and armaments production was increased or at least maintained to a certain extent until the end of the war, although there were increasingly serious bottlenecks in the supply of certain raw materials. For example, the

Influence of the State

Above: Fokker fighter E.III 410/15 after repair (M14) (1915).

production index of the iron and steel industry group fell from 100 in 1913 to 53 in 1918, that of the mining industry from 100 to 93 and that of the textile industry, a mixed industry, from 100 to 17. The aviation industry was particularly affected by the shortage of suitable types of wood, fibrous materials for clothing, certain metals, raw rubber, lubricants and fuels. As the war progressed, inferior substitutes that impaired the quality and operational readiness of the aircraft had to be used more and more frequently, despite priority being given to the allocation of materials in aircraft construction, first for training aircraft at home, then also for front-line aircraft: brittle iron instead of soft metals for tanks, pipes and accessories, wooden disks instead of rubber-tyred wheels or lower-quality fuel mixtures for flying in and for training flights.

Shortage of Workforce

In addition to the shortage of materials, bottlenecks in the supply of sufficient manpower were a limiting factor in the armaments industry. High losses of fallen, wounded and sick soldiers and the repeated army reinforcements caused the demand for front-line soldiers to rise steadily, and more and more trained specialists were called up. A department in the War Ministry and later in the War Office was entrusted with issues relating to the release of workers and the supply of manpower, but the General Staff and OHL saw the priority as securing and increasing the fighting strength of the field army and wanted to call up as many men as possible fit for military service for front-line service; they regarded deferments as a weakening of the fighting troops that should be avoided as far as possible.

In the course of the Hindenburg Program, the number of workers deferred temporarily rose to around 200,000 by mid-1917, but then the OHL began to call up an increasing number of men for military service again.

The aviation industry enjoyed a privileged position in the release of its skilled workers and the allocation of additional workers since August 1914 as a result of being given priority, and the total number of employees in aircraft and aircraft engine construction continued to rise steadily during the war years, with some fluctuations. Bottlenecks occurred in the supply of qualified skilled workers: Aircraft represented complicated weapon systems not only in operation but also in production, and the complex work processes in airframe and especially engine construction required trained skilled workers who could hardly be adequately replaced by unskilled or semi-skilled workers. Although the use of machines that could also be operated by workers without specialist training made it possible to use a larger contingent of unskilled or semi-skilled workers, men, women and young people, their proportion in aircraft production in Germany does not appear to have been particularly high - Morrow[55] (Refer to Morrow, John H.: *German Airpower in World War I.*) puts the figure at no more than 25% of the total workforce. In the final years of the war, the strain on workers due to long working hours and the poor food situation as a result of the Allied blockade continued to increase and also led to repeated work stoppages in the aviation industry, despite the relatively high wages and a privileged position in the food rationing system: The workers, whose position vis-à-vis the employers was strengthened by the lack of personnel, demanded an adjustment of wages to the constantly

Development of the Aviation Industry During the First World War

Left: Final assembly of Pfalz D.XII fighters with Daimler D.IIIa engines.

rising cost of living and a reduction in working hours. The inadequate supply of food led to further stoppages, with workers repeatedly calling in sick to make hoarding trips to the countryside to obtain food.

Although the aviation industry was not able to fully meet the ambitious quantitative targets of the Hindenburg program, the support measures contributed to a further increase in production. With the **America Program**, the next expansion program of the air forces began in the spring of 1917: The OHL planned a major offensive on the Western Front for the spring of 1918, which was intended to force a decision on the war before the armaments machinery of the United States of America, which had entered the war in April 1917 with its almost inexhaustible economic resources, got into full swing and finally shifted the balance of power in favour of the Allies. Although the new enemy had huge human and material resources at its disposal, the small arms industry, which nevertheless promised immense potential, first had to be mobilized and an army of millions raised, trained, equipped and transported to the European theater of war. As the air forces played an elementary role in the German attack plans to support the breakthrough attempts of the ground troops and a significant increase in the Allied air forces was also expected, in June 1917 the Kogenluft demanded a doubling of the industry's output to 2,000 aircraft and 2,500 engines per month by March of the following year in order to reinforce existing units and set up 57 additional aviation units. The War Ministry objected to these ambitious targets and the figures had to be reduced to a maximum of 1,600 aircraft and 1,800 engines, as the huge material and personnel requirements of such an enormous expansion threatened to jeopardize other armaments projects that were considered equally important. Nevertheless, the attempt to meet the military's planned figures could only be made at the expense of other armaments projects: As a weapon classified as crucial to the war effort, the OHL finally gave priority to the aircraft over the tank, with only the construction of submarines taking precedence over aircraft production. All available forces were to be concentrated on preparing for the spring offensive, which was regarded as decisive:

"I have repeatedly pointed out to the aircraft industry, both verbally and in writing, that for military reasons the industry's ability to supply aircraft and aircraft spare parts must be strained to the utmost," said Major Felix Wagenführ of the Luftwaffe in a circular letter to the aircraft factories in February 1918.

However, the shortage of raw materials and skilled labor could not be fully compensated for by the priority given to the supply of materials and personnel. Although the quantitative targets of the America program were not achieved within the planned time frame - even in the best month, March 1918, the output of 1,360 aircraft remained below the required 1,600 aircraft - a considerable increase in production was once again achieved thanks to the government measures: while the monthly output in autumn 1917 averaged 900 aircraft, it rose to 1,150 aircraft in the first seven months of 1918. In the summer months of 1918, the aviation industry then reached its peak performance, with some fluctuations: 1,500 aircraft were completed in July and no fewer than 2,195 in October A similar development can also be observed in engine production, where production had increased to 1,380 units per month in the first half of 1918.

The high value placed on aircraft by the military

Influence of the State

Right: Gotha G.I with two counter-rotating Benz Bz.III and Reschke propellers, 2 kg bombs on the nose, and one 10 kg bomb under the lower wing.

leadership guaranteed aircraft construction priority in the supply of materials and personnel. In order to ensure the most efficient use of available resources, to increase quantity and quality and to assert its interests, the leadership of the air forces took a series of controlling, directing and coordinating measures: Continuing the development that had already begun in the pre-war period, the military further expanded its direct influence on the companies during the war. From the beginning of the war, officers in charge of quality control supervised the production of aircraft and engines in the factories and monitored compliance with the construction and delivery regulations issued by Idflieg, which laid down guidelines for design and production. In order to better control the construction process and to be able to intervene directly in all stages of the development and production process, Idflieg set up permanent military construction supervisors as organs of the Central Acceptance Commission in the larger factories from 1916. In addition to monitoring the allocation of raw materials and the construction of aircraft and engines, their tasks also included the acceptance of finished aircraft and mediation between employers and employees, or if problems arose that threatened production. The number of construction supervisors, who were assigned not only to important airframe and engine manufacturers but also to companies in the accessories industry, grew from 19 to almost 50 by the summer of 1918.

In order to secure the flow of production, the military administration intervened in internal industrial matters that had previously been considered sacrosanct for state authorities, in the procurement of labor and raw materials, but also in employer-employee relations and the regulation of employment relationships within companies: For example,

Above: Schütte-Lanz Dr.I experimental triplane fighter (1918).

the War Ministry encouraged collective negotiations on wage issues involving the trade unions, and officers from the building inspectorates acted as arbitrators between employers and employees in conflicts over wage issues. The procurement authorities of the air force, who wanted to prevent disruptions to production at all costs due to the increasing demand for aircraft, repeatedly supported the wage demands of employees in labor disputes and urged companies to raise wages to compensate for the increased cost of living, regardless of the possible effects on other branches of industry. As in other belligerent superpowers, it can be observed that the deprivations caused by the war, coupled with increasing demands for performance, fueled employee dissatisfaction and the state supported their demands on employers to ensure the continuation of the war effort. An industrialized war could not be waged without the cooperation of organized labour, so military and civilian authorities attempted to come to terms with the workers.

In the case of Junkers, the command of the air force even intervened in the company's independence in order to secure production: Since the Junkers factories, which were inexperienced in aircraft production and relatively financially weak, were not trusted to mass-produce the J4 (J.I) infantry aircraft ordered in large numbers on their own, the Idflieg urged Junkers to cooperate with a large, financially strong aircraft factory, Fokker, in order to combine technical capabilities with the ability to mass-produce. Junkers-Fokker-Werke AG (Ifa), which was founded in October 1917, existed until April 1919 and produced around 320 aircraft. Serious tensions soon arose between the two owners: Hugo Junkers suspected that Fokker was primarily committed to the prototypes of his own company in Schwerin, which continued to exist independently, while Anthony Fokker accused Junkers of stubbornness, as he would not recognize the wartime primacy of producing large quantities as quickly as possible over a technical revolution such as all-metal construction.

The military authorities attempted to counteract the factors limiting quantitative development, in particular the shortage of personnel and raw materials, by standardizing, concentrating and rationalizing the industry: Standardization on a limited number of high-performance aircraft types and their licensed construction by a large number of companies was intended to increase production figures, make better use of the available forces and also simplify logistics and operations at the front. The incentive for further technical development, which should by no means be neglected, should lie in the fact that the model with the best performance should be prescribed to the other factories for reproduction.

At the production level, standardization was intended to facilitate reproduction and series production, and from the spring of 1917 Idflieg also attempted to enforce generally applicable industrial standards in aircraft production. And finally, a concentration of the industry was to ensure the most efficient use of raw materials, the available quantity of which was decreasing. Idflieg intended to call on increasingly

Above: Although designated a G type, which covered twin-engined aircraft, the Roland G I was in fact a single-engined aircraft with twin pusher airscrews. The engine was buried in the fuselage on the centre of gravity and the airscrews driven through a system of gears and shafts. Only the single prototype, which carried a crew of two, was built. Armament, one flexible Parabellum machine-gun.

efficient, large companies capable of mass production for production orders, which could fulfill the contracts quickly, and to assign only repair and maintenance work to factories that did not meet these requirements. By the spring of 1917, a group of companies considered to be productive had emerged, which were given preference in the awarding of large contracts and thus also in the allocation of materials. If the companies proved themselves, they received further orders, otherwise their quota was reduced and the contracts were awarded to other factories. Not only the airframe and engine manufacturers were affected by these measures, but also the accessories industry: Idflieg selected a number of companies for army deliveries and issued instructions to the aircraft and engine factories that parts for military aircraft could only be purchased from approved companies. Rationalization and centralized control, which made it easier to redistribute resources from less productive to more efficient companies, played a decisive role in ensuring that aircraft production could be maintained despite the bottlenecks that arose and that output could even be increased further until the late summer of 1918.

In order to facilitate licensed production and make technical developments that were considered important for improving performance more easily accessible to all companies, the military authorities attempted to relax patent protection in their favor. At the beginning of the war, the eight largest aircraft companies had agreed that patents of lesser importance could be used by the military without a license and that any disagreements that arose would be settled by a parliamentary commission. After the strong growth of the aviation industry seemed to make a clear clarification of the patent issue necessary, a conference was convened in January 1916: In contrast to the established aircraft companies, the military insisted on license-free use of the patents, which was only a small price to pay for the large wartime production orders and the large number of technical suggestions from the field, the suggestions for improvement from the front-line units, which would be passed on to the industry's design offices via the Central Acceptance Commission (ZAK) and the construction supervisors. However, in six months of negotiations, the industry was able to assert its interests and achieve a broad retention of the patent laws. The parliamentary commission recognized the companies' claims as legitimate, and in return the industry agreed not to make any further claims during the war under certain conditions. The agreement of August 1914 remained in force, otherwise the companies building under license for the military had to pay a fee to the patent holder and the courts were to decide on disputed issues. In the war situation, however, the legal protection itself was unstable - following a modification of the patent laws by the Federal Council, the War Ministry was able to prohibit the publication of patent

Development of the Aviation Industry During the First World War

Above: Engine test stand of the Deutschen Versuchsanstalt für Luftfahrt (DVL) in Adlershof.

applications classified as important to the war effort, which enabled the military to use an invention and still refuse to pay fees, as the applicant could not assert any claims before the patent was published.

In the final years of the war in particular, the Allies together had a considerable numerical superiority in front-line aircraft, supported by far greater capacities in aircraft construction and the ability to draw on larger stocks of raw materials. The German air forces attempted to counter this quantitative superiority, which became increasingly oppressive from the summer of 1918, with aircraft of better technical quality and superior performance: In June 1918, the Kogenluft shifted the focus of the expansion program, which could not be fulfilled quantitatively, from increasing the numbers of the air force to increasing combat power by introducing new fighter aircraft with superior performance. Idflieg provided the companies with technical guidelines and held three comparative flights with different fighters in Adlershof in order to select the best models for series production. However, the differences in performance between the best German and Allied designs were too small to compensate for the numerical inferiority of the German air force through quality - it became apparent that the quantity of aircraft, even of inferior quality, was a significant factor in air warfare.

As private industry had hardly any research and testing facilities, the military and civilian authorities continued to expand state-controlled aviation research during the war in order to promote aviation technology; recourse to reliable scientific findings was intended to enable companies to design new generations of aircraft with higher performance: The DVL, which finally relinquished its status as an independent civilian institution in August 1914, was able to greatly expand its facilities in Adlershof, which were shared with the Fliegertruppe (PuW) testing institute and shipyard and later the army's Flugzeugmeisterei, in cooperation with the military authorities. The War Ministry and the Reich Navy Office also contributed financially to the expansion of the Göttingen Model Testing Institute, which in return undertook to give priority to military research contracts. In January 1917, the Idflieg set up the Scientific Information Office for Aviation (WAF), which was affiliated to the Aircraft Master's Office and whose tasks consisted of "opening up the scientific labor market of all eligible research institutes and specialists, acting as an intermediary and, if necessary, keeping it fully occupied with suitable orders from the army and navy administration for the purpose of its fullest utilization. Unnecessary and parallel work must be avoided at all costs during the war."

In order to improve the flow of information for the benefit of the further development of aviation technology, the WAF, as a central office, was also to collect relevant scientific and technical information obtained from the research orders placed and other sources, including the evaluation of captured aircraft, engines and instruments, and distribute it to the authorized bodies in research, the military and industry by means of technical reports. Due to the special requirements of the war situation, most aircraft companies agreed to waive the protection period stipulated in the WAF's rules of procedure for the publication of results from research and test work commissioned by companies in the Technical Reports.

Summary

The major European states expected a short war with limited use of aircraft. The demands on the military and economy only took on different dimensions with the actual development of the war. None of the states were prepared for a four-year, material-intensive war of position. It now became crucial for the success of the war effort to mobilize economic and industrial capacities in line with the changing requirements and put them at the service of the war effort.

Through supportive, coordinating and controlling measures, the authorities attempted to create an environment that promoted a conversion and an increase in the production of armaments. Although Germany's aviation industry, like that of Great Britain, was unable to meet the military's quantitative targets, it increased its capacity and output of aircraft many times over between August 1914 and November 1918. The military authorities' requirements reflected the military planners' top priority, namely the

Influence of the State

Right: Junkers wind tunnel in Dessau (ca. 1917).

Left: Wind tunnel of the Göttingen Aerodynamic Research Institute (ca. 1918).

greatest possible reinforcement of the field army by supplying as much material as possible. This was done regardless of the fact that in order to achieve the military objectives and conduct the war victoriously, a battle for resources and power on the "home front" with other departments was unavoidable. As far as the use of the new weapon of aircraft and aerial warfare was concerned, there was still no experience; the decision-makers in the civil administration and the military had to plan new developments for an open future. In order to achieve the war aims and to be adequately equipped against the enemy, the material requirements tended to be set higher, sometimes disregarding a realistic assessment of the available resources. Thus, in pursuit of its goal of a victorious peace, the 3rd OHL drew up ambitious expansion programs that were not based on a sober assessment of the resources available in Germany and thus already bore the risk of numerical failure.

In absolute terms, the aviation industry expanded considerably during the war, primarily through the expansion of existing pioneering companies and the diversification of companies from outside the industry into aircraft construction, which was largely seen as a temporary measure for the duration of the war. Less importance was attached to the founding of new aircraft companies during the war years and the attempt made in Great Britain, but not in Germany, to produce in state-owned factories. As a result of the procurement policy of the pre-war period and the structural development, there were differences in emphasis in the development in Great Britain and the German

Development of the Aviation Industry During the First World War

Left: Halberstadt D.I prototype with front radiator in contrast to D.II.

Reich: in Germany, a group of large aircraft companies had emerged by the summer of 1914 as a result of orders from the military for larger quantities, which, together with some large companies that diversified into aircraft construction during the war, were essentially responsible for the expansion of production capacity. With state support, the pioneering companies in Great Britain were also able to expand, whose production capacities had remained predominantly low during the build-up phase of the air forces due to smaller military orders, but the expansion of the industrial base by bringing in a larger number of companies from outside the industry to build aircraft based on designs by the state-owned Royal Aircraft Factory and private companies was of greater importance for the increase in production. The greater wear and tear on material during the war and the increase in the value attached to air support for the ground troops demanded far greater production than the sales potential in peacetime; at the beginning of the war, the aviation industries were geared towards a lower demand based on the military's pre-war planning and were hardly sufficiently prepared for the mass production of such high quantities as the military now demanded in the changing situation. In addition, the range of aircraft categories required increased with the expansion of the air force's remit; by the end of the war, it

Above: The robustness of German aircraft is clearly demonstrated by this Hannover CL.II. The fabric was partially torn off.

Above: Hansa-Brandenburg FB Flying boat #511.

ranged from fighter monoplanes, reconnaissance aircraft and armored infantry aircraft to heavy long-range bombers and flying boats. While circumstances resulting from the war situation, such as the shortage of skilled labor and available raw materials, hindered the further expansion of the aviation industry to a certain extent, the special situation also favored an accelerated development of capacities: Economic self-organization and technological modernization of the industry were important factors that contributed to the expansion of aircraft production during the war years,

Development of the Aviation Industry During the First World War

Left: Cover of the magazine "Motor", issue Jan./Feb. 1918.

but major impetus came from the state, which saw the availability of large numbers of aircraft as necessary for the success of the war effort: The state promoted the aviation industry through greater profit opportunities due to increased demand or direct support such as the cost-reimbursement contracts and by creating favorable conditions, by providing labor and materials, even at the expense of other branches of the armaments industry due to the high priority assigned to aircraft construction, and attempts to pacify conflicts between employers and employees through coercive measures such as arbitration. However, this also shows that the state extended its control and influence over industry beyond technical issues such as performance requirements and quality assurance and intervened deeply in the internal affairs of industry, in the management of raw materials, entrepreneurial independence, the employer-employee relationship and wage issues, in favor of securing the flow of production, which was elementary for the war effort. The army became the catalyst of a revolution from above, for opportunistic motives, in order to prevent a revolution from

Above: Light combat aircraft Halberstadt CL.II (C.15459/17) in the Berlin Aviation Collection (opening: June 20, 1936).

Above: A Halb CL.IV(Rol) with Daimler D.IIIa engine built under license by Luftfahrzeug-Gesellschaft (LFG).

Development of the Aviation Industry During the First World War

Above: Albatros D.III with ground crew in front of an airship hangar.

below and to secure the production of war goods. In terms of the quantitative development of the industry and the influence of state bodies, the beginning of the war did not represent a radical turning point, but rather an intensification of tendencies that had already characterized the aviation industry in the pre-war period. In the aviation industry, the increase in production in the environment created by the state can be observed on two levels: On the one hand, the dimensions of the factories increased sharply, but on the other hand, the total number of aircraft manufacturers called upon for army and navy orders also grew – during the war, around 35 new companies were established, and in the fall of 1918, a total of around 70 companies were manufacturing airframes for the armed forces. Some companies, including pioneering firms such as Hans Grade Flieger-Werke or Flugmaschinenwerke Gustav Otto, had to cease operations during the war or were taken over by more successful companies, as they were unable to meet the military's requirements for efficient mass production and so orders failed to materialize, but the basic development trend was an expansion of the industry between the beginning and end of the war. Large orders went to the major army suppliers of the pre-war period, but as a result of the increasing demand, other companies were also included and received repair and production contracts, existing factories that had previously hardly been considered for military orders, or newly founded companies. Based on the expectation that the growing demand would make it possible to participate in the aircraft construction business, entrepreneurs founded a number of new companies, especially in the first years of the war: Around 20 factories were established in 1914 and 1915, and a further 14 by the end of the war. Initially, new companies often copied tried-and-tested types from other factories under license, the prospects of more lucrative contracts for their own developments being considered favourable due to the great demand, but they also encouraged companies that had not previously been involved in construction to design aircraft themselves, albeit mostly with little commercial success. A representative example is Mercur-Flugzeugbau GmbH, which was founded at Johannisthal airfield in 1915. Initially, the company was only used by the military for repairs, but soon it was also able to win orders for the licensed production of B and C aircraft based on Albatros and Rumpler designs. By 1918, the number of employees had risen from 70 to over 1,400 and the number of aircraft delivered from 370 in 1916 to 600 in 1918.

In the last two years of the war, one of the company's engineers, Fritz Hildebrandt, developed several fighter monoplanes, some of which were built as prototypes but rejected by the Luftwaffe. Independent companies made up only a part of the newly founded companies, and there were also a number of other companies from outside the industry that set up aircraft construction departments: Daimler Motoren-Gesellschaft, Hannoversche Waggonfabrik AG (HAWA) and Waggonfabrik Linke-Hofmann were among the companies that diversified into aircraft production;

Above: Junkers J4 (military designation Junk J.I) shortly before completion (Dessau 1917).

Siemens-Schuckert-Werke, which had started producing grenades and fuses at the beginning of the war, reactivated its aircraft construction department, which had been closed in 1911, in October 1914 as part of the expansion of war production. Luftschiffbau Schütte-Lanz founded an aircraft department in 1914, and the Zeppelin Group participated in the manufacture of "heavier-than-air" aircraft through Zeppelin Werk Lindau GmbH, which emerged from the Do department under Claude Dornier, Zeppelin Werk Staaken GmbH and the participation in Flugzeugbau Friedrichshafen GmbH, From fighter and naval combat aircraft to giant airplanes and long-range flying boats Since the establishment of large and efficient production facilities required capital and know-how for efficient series production, the military authorities promoted the entry of large corporations. Due to their solid economic basis, they were considered capable of easily providing the necessary funds for the operation, expansion and modernization of the facilities, and they were also expected to be more stable in the face of fluctuations in the order situation. In addition, industries such as wagon construction already had suitable machinery, skilled workers trained in wood and metalworking and the know-how to manufacture in large quantities.

The emergence of new factories broadened the industrial base, while the dimensions and production capacity of the individual plants grew through the expansion of existing facilities or the establishment of branch plants. In October 1916, for example, the Berlin Rumpler-Werke and the August Riedinger Ballon-Fabrik AG in Augsburg founded the Bayerische Rumpler-Werke AG, which carried out repairs and built Rumpler C aircraft, LVG set up a branch company in Köslin for aircraft production and repair, and Albatroswerke GmbH from Johannisthal opened branch factories near the Müggelsee near Berlin, in Schneidemühl and in Warsaw between 1914 and 1916. In general, the number of employees and the floor space of the construction and production facilities increased sharply during the war, especially in the first years of the war, until the shortages that occurred despite the government measures curbed the upward trend. However, the number of employees as an index of growth does not necessarily say anything about the rate of increase in production; an increasing proportion of unskilled or semi-skilled workers in a growing total number of employees could certainly have a negative impact on productivity. The number of aircraft delivered fluctuated, but the basic trend that emerged was an increase in the monthly output of airframes. Looking at the delivery figures of the individual plants, it is important to note which aircraft

Development of the Aviation Industry During the First World War

Above: Flugzeugbau Friedrichshafen FF38, here the G.II 6xx/16 built at Daimler DMG, Section Flugzeugbau.

categories the companies produced: C and D aircraft were easier and quicker to produce in larger quantities and were also ordered in larger numbers than G or R aircraft. The total production of AEG's aircraft construction department, for example, consisted of around two thirds C and J aircraft and one third G aircraft, while Albatros produced almost exclusively B, C and D aircraft.

The pre-war military orders had already created a financial basis for the expansion of the industry. During the war years, the expectation of a steep rise in demand, not least in connection with the Hindenburg and America programs, as well as the pressure of competition from other expanding companies, led the owners to invest large sums in the companies for further expansion or conversion of production to new models, with the profits from the large orders helping to provide the necessary capital. Gothaer Waggonfabrik, for example, tripled the share capital of its aircraft construction department to three million marks in order to be able to start manufacturing G-planes. Independent aircraft companies that were not backed by large corporations had to find other ways to strengthen their financial base in order to expand their production capacities: Rumpler converted its company, which had been entered in the commercial register as a GmbH in the summer of 1914, into a stock corporation in September 1917 and was able to increase its share capital to 3.5 million gold marks due to the favorable prospects for aircraft construction. In August 1918, the company spokesmen then announced that the company had made a profit of 1.02 million marks in its first year as a public limited company and would pay a dividend of 12%. More detailed information on the economic development of the aviation industry is hard to find, but one indicator that successful companies were making a profit is the development of dividends: Automobil und Aviatik AG, which was one of the largest suppliers to the army in the pre-war and wartime period, paid a dividend of 8% in 1913, 20% the following year and then 25-30% in 1917. As a result of the increase in wages for skilled workers and raw material prices not controlled by the state, the production costs for airplanes rose, but the military ultimately had to agree to an increase in the prices to be paid for the airplanes demanded by the companies. In the first years of the war, the War Ministry maintained a strict pricing policy and only granted small price increases, as it was believed that the larger contracts alone would bring an appropriate increase in profits. However, when the price of raw materials and the wages of skilled workers began to rise sharply due to a shortage and increasing demand, the procurement authorities also had to accept rising prices - the importance of air support for the war effort due to the growing number of aircraft and the dependence on private industry strengthened the position of companies in relation to the military to a certain extent during the war.

At the end of the war, the aviation industry was characterized by large companies with high production capacity: successful companies such as Albatros, Fokker, Rumpler, LVG or Flugzeugbau Friedrichshafen in the field of naval aircraft, which usually had fewer than 500 employees

Influence of the State

at the beginning of the war, employed an average of 2,000 to 3,000 people in 1918, and the size of the production facilities had multiplied. At the beginning of the war, the total workforce in the airframe, engine and accessories industry amounted to around 6,000 employees, whereas in the last year of the war around 40,000 to 50,000 people were probably employed in aircraft construction, plus a further 100,000 in engine production and the accessories industry. From August to December 1914, the industry had delivered a total of around 700 aircraft, compared to around 17,000 from January to November 1918. The total number of aircraft produced in Germany during the First World War and accepted by the military varies in the literature, ranging from 45,704 to 47,931.

This book cannot provide a conclusive, all-encompassing assessment of the various data, but it should encourage readers to exhaust all possibilities in order to ultimately verify one or other of the figures. However, the overall result will not change much. Enormous efforts were made in all countries involved in the war to ensure that a young technology such as aircraft construction could develop to a high level in a very short time.

Endnotes

1. Viktor Stoeffler (* June 9, 1887 - † July 1, 1947), license number 174, was a pioneer of French aviation. He was a test pilot at Aviatik in Mulhouse, Freibourg (1912-1913) and its later factory director, then in Leipzig. (1914-1918).
2. Ernst German Schlegel (* June 21, 1882 in Konstanz; † January 4, 1976 ibid.) was a German aviation pioneer, aircraft designer and engineer. In 1909, he moved to Mainz to attend the engineering school there. Based on his own designs and plans, he constructed the "Schlegel-Züst flying machine" together with his Swiss friend Robert Züst. Schlegel was a mail pilot, test pilot, water pilot, and then a wartime pilot in World War I with numerous missions. He was promoted to lieutenant and received the Iron Cross 1st and 2nd class. Later he was a senior technical employee at Rumpler-Flugzeugwerke in Berlin and in 1917 was transferred by the army administration to the Pfalz aircraft factory in Speyer as chief engineer.
3. Karl Christian Maximilian Caspar (* August 4, 1883 in Netra (Hesse-Nassau); † June 2, 1954 in Frankfurt-Höchst) was a German pilot, aircraft manufacturer and lawyer. He gained notoriety, especially in the 1920s, for the types of aircraft developed and built by Caspar-Werke. As an airplane pilot who took his exam before the First World War, he is counted among the Old Eagles.
4. Robert Thelen (* March 23, 1884 in Nuremberg; † February 23, 1968 in Berlin-Hirschgarten) was a German engineer, pilot and aviation pioneer. During World War I, Thelen was employed by Albatros Flugzeugwerke as a test pilot and designer; at the same time, Ernst Heinkel was responsible for designing bomber aircraft. Among Thelen's designs was the Albatros D.III, one of the Air Force's most widely used fighter aircraft. As a test pilot, he flew the Albatros seaplanes built at Friedrichshagen.

Above: The all-metal Zeppelin D.I, with its fully semi-monocoque construction, had the most sophisticated structure of any WWI aircraft. Designed by Claude Dornier, its structural technology is still being used.

Above: AEG R-aircraft construction hangar.

Above: AGO Flugzeugwerke, Johannisthal.

Above: Aviatik factory, Johannisthal.

Above: Hannover factory, Johannisthal.

Above: Roland D.VIb at LFG firing range.

Above: LVG factory, Johannisthal.

Above: Pfalz factory, 1919.

Above: Rumpler aircraft ready for shipment..

4. Development of German Aircraft Factories During the First World War

4.0 General

In the following chapters, all major aircraft, aircraft engine and propeller manufacturers are presented. The production figures were taken from the report of the Interallied Monitoring Commission of Germany in the field of aviation (ILÜK). In accordance with the Treaty of Versailles, this commission monitored all major manufacturers of aircraft, engines, propellers, etc. involved in wartime production. They also controlled the aviation material still in stock in the factories after the end of the war and supervised their destruction.

Number of Aircraft Built in Germany									
Group	1911	1912	1913	1914	1915	1916	1917	1918	Total
A	11	60	168	294	13	22	–	–	568
B	13	76	278	1,054	1,312	440	2,993	25	6,191
C					2,674	2,726	10,337	7,320	25,057
D					1	2,129	4,945	5,132	12,207
Dr							338	1	339
E					347	300	–	381	1,028
G, R					185	465	589	789	2,028
J							450	463	913
N						100	94	10	204
S								2	2
Total	24	136	446	1,348	4,532	8,128	19,746	14,123	48,537

Aircraft Built in Germany by Category				
Category	Configuration	Intended Role	No. Engines	Quantity
A	Monoplane, Taube	Training and observation aircraft	1	329
B	Biplane	Training and observation aircraft	1	5824
C	Biplane	Reconnaissance & artillery spotting	1	25,057
CL	Mono & Biplane	Escort & attack aircraft	1	
CLS	Mono & Biplane	Light escort & attack aircraft	1	
D	Mono & Biplane	Single-seat fighter	1	12,546
DJ	Mono & Biplane	Armored single-seat fighter	1	
Dr	Triplane	Single-seat fighter	1	
E	Monoplane	Single-seat fighter	1	1,028
J	Biplane	Armored infantry cooperation	1	913
N, CN	Biplane	Single-engine night bomber	1	204
G	Biplane & Triplane	Bomber	2	2028
GL	Biplane & Triplane	Lighter bomber	2	
L	Biplane	Bomber	3	2
R	Biplane	Heavy bomber	3–6	
Total				47,931
See also Gilles - Technical report translation and internet material				

Aircraft Production in Germany According to Militärarchiv Freiburg (RL2-IV-303)

Manufacturer	Alias	1914	1915	1916	1917	1918	Total
Ago Flugzeugwerke GmbH, München u. Johannisthal	Ago	60	120	300	600	800	1880
Otto-Werke GmbH	Ot						~200
Albatros-Gesellschaft für Flugzeugunternehmungen	Alb	338	846	1011	2097	1950	6242
Ostdeutsche Albatroswerke GmbH, Schneidemühl	Oaw	40	400	400	400	400	1640
Albatros, Militärwerkstätten, Warschau	Refla						200
Allgemeine Elektrizitätswerke GmbH	Aeg	20	91	220	615	454	1400
Automobil- und Aviatik AG	Av	165	316	367	852	1100	2800
Bayerische Flugzeugwerke AG	Bay			389	1039		~2600
Bayerische Rumpler-Werke AG	Bayru				46	271	317
Daimler Motoren-Gesellschaft	Daim						~296 (220)
Deutsche Flugzeugwerke GmbH	Dfw	93	79	99	820	1100	2191
Euler-Werke	Eul	53	51	33	130	55	322
Flugzeugwerke Albert Rinne	Rin						~600
Fokker Flugzeugwerke GmbH	Fok	32	260	675	798	1500	3330
Germania Flugzeugwerke GmbH	Germ						300
Gothaer Waggon-Fabrik AG	Go						>529
Halberstädter Flugzeugwerke GmbH	Halb	56	175	320	550	820	1921
Hannoversche Waggonfabrik AG	Han						
Hansa und Brandenburgische Flugzeugwerke GmbH	Brand						~600

Aircraft Production in Germany According to Militärarchiv Freiburg (RL2-IV-303) (continued)

Manufacturer	Alias	1914	1915	1916	1917	1918	Total
Hanseatische Flugzeugwerke AG (Karl Caspar)	Hansa				200	93	293
Junkers-Fokker-Werke AG	Jfa (Junk)						~290
Kondor-Flugzeugwerke	Kond						~500
Linke-Hoffmann Werke AG	Li			12	120	186	318
Luft-Fahrzeug-Gesellschaft mbH	Rol (Lfg)	15	73	512	893	1366	2859
Luftfahrzeugbau Schütte-Lanz	Schül						~1000
Luftverkehrsgesellschaft mbH	Lvg	300	1020	720	1680	1920	5640
Märkische Flugzeugwerft GmbH	Mark						~500
Mercur Flugzeugbau GmbH	Merc			370	450	600	1420
Pfalz Flugzeug-Werke GmbH	Pfal	14	94	207	793	~1000	~2200
Rumpler-Werke GmbH (später AG)	Ru	109	210	486	901	1400	3106
Sablatnig-Flugzeugbau GmbH	Sab				30	30	60
Siemens-Schuckert-Werke GmbH	Ssw			250	350	380	980
Zeppelinwerk Lindau	Do						~25
Zeppelin Staaken	Staak		2	4	6-7	3-4	~20-24
Flugzeugbau Friedrichshafen GmbH	Fdh	36	87	187	433	574	1320
Luftschiffbau Zeppelin	Zep						20
Total		1,348	4,532	8,182	19,746	14,123	47,931

Note 1: The numbers vary from source to source.
Note 2: Numbers in red are according to own findings.

4.1 AGO-Flugzeugwerke GmbH and Otto Flugzeugwerke

Company History

Group	1909	1910	1911	1912	1913	1914	1915	1916	1917	1918	1919...
Munich Puchheim		Academy of Aviation (flight school)	→	→	→	→	→	→	→	→	→
Munich Oberwiesenfeld	Aeroplane construction Otto & Alberti	Aeroplane construction Otto & Alberti									
		Bayerische Auto-Garagen GmbH B.A.G.	→	→	→						
	Aeroplanbau Otto & Alberti	Aeroplanbau Otto & Alberti	Aeroplanbau Otto & Alberti, ab 15. März 1911: Flugmaschinenwerke Gustav Otto	→	→	→					
			Ago Aeroplanbau Gustav Otto	→	→	→					
			A.G.O. Aeromotor Gustav Otto	→	→	→					
							Otto Flugzeugwerke und Maschinenfabrik GmbH, Otto-Werke GmbH (Schleißheimer Str. 135) (bestanden formal bis zum 03. September 1937)	→	→	→	→
Berlin-Johannisthal			Aviatiker Gustav Otto, AGO Fluggesellschaft mbH	→	→	→					
							AGO-Flugzeugwerke GmbH	→	→	→	→
Oschersleben/Bode									Aktiengesellschaft Otto	→	→

After completing his studies in 1909, Gustav Otto initially founded three companies in Munich: "Bayerische Auto-Garage GmbH" (B.A.G.), the "Akademie für Aviatik" and "Aeroplanbau Otto & Alberti/Gustav Otto Flugmaschinenwerke", although it should be noted that Gustav Otto initially devoted himself to aviation technology purely out of sporting interest. In 1909/10, he took over the German representation of Blériot and the biplane manufacturer "Aviatik GmbH" from Mulhouse (Alsace).

The "Academy of Aviation" newly founded by Otto prompted him to move from Munich to Puchheim and, in the summer of 1910, to become the 34th German pilot to take his pilot's certificate on an aviation flying machine with a 50 HP Argus engine. Otto then trained various pilots at his academy on the Puchheim airfield near Munich, including Ernst Udet (1896-1941) in 1914.

His "Aeroplanbau Otto & Alberti" was initially set up in 1909 at Munich's Oberwiesenfeld, at that time a drill ground which G. Otto was allowed to use as an airfield by agreement with the Bavarian military authorities. In 1910, both opened a branch of the company in Puchheim, the seat of his Academy of Aviation. Puchheim had a newly opened airfield and thus offered quite good conditions for the repair and manufacture of flying machines. 40 employees and technical staff were initially only responsible for the production of replicas of the Blériot monoplanes. With 40 employees produced mainly replicas of the Bleriot monoplane. The first Bleriot replica, the "Otto No. 1" was christened the "Sperber" (Sparrowhawk).

The Bleriot monoplane was followed by replicas by Henry Farman. Unlike the Bleriot, the Farman biplanes had a pusher propeller. The grid fuselages were not covered on the Bleriot or the Farman airplanes.

As late as 1910, Otto had his own aircraft designed by Gabriel Letsch on the basis of the French prototypes. These first airplanes already had to meet important criteria that were expected of military airplanes, such as quick disassembly, fast assembly and good loading characteristics. The flight characteristics were relatively uncomplicated, so that these grid fuselage biplanes were gladly used in flight schools. The crew consisted of a passenger and

AGO & Otto

> **A. G. O.**
> **Aeroplanbau Otto & Alberti**
> (Inh.: Gustav Otto)
> München   Karlstr. 72.

> Fertigt die erfolgreichen
> **Doppeldecker Typ A. G. O. 1**
> 1. tägl. Dauerpreis Berlin 1910
> 3. grosser Flugpreis von Magdeburg 1910
> **Passagierzweidecker Typ A. G. O. 2**
> **Passagiereindecker Typ A. G. O. 3**
> Leichteste Construction aus Holzrohr.   Flugunterricht auf 5 Flugapparaten.

Above: AGO-Advertising 1911.

the pilot, who sat directly in front of the engine. This was also an unfavorable aspect of the design: Since the Otto fuselage biplane was not yet equipped with a so-called shock landing gear, it could happen that the nose of the aircraft made contact with the ground in the event of an ungentle landing. In such a case, there was a risk that the pilot and passenger would be killed by the engine. However, there were large skids between the pairs of wheels of the landing gear, which were able to dampen forward pitching movements.[1]

In 1910, B.A.G. was restructured in Munich into an aircraft repair shop for Blériot monoplanes, and Aeroplanbau Gustav Otto was founded in Puchheim. For the first time, this company bore the three letters AGO as its abbreviation. The Puchheim location with the directly adjacent airfield now proved to be too costly and too remote. The uneven nature of the field also led to considerations of moving the location of the plant and the academy to Munich. Nevertheless, Otto maintained a presence in Puchheim. The flying school used Otto biplanes and Otto aircraft were also set up for the flying meetings. In the spring of 1911, Otto moved its operations back to a newly built wooden shed near the Oberwiesenfeld parade ground in Munich, the old location that was to prove formative for Bavarian industry: Schleißheimer Strasse on Oberwiesenfeld in Munich. Thus, Aeroplanbau Otto & Alberti and AGO, Aeroplanbau Gustav Otto were located in Oberwiesenfeld at the same time.

He repaired Blériot airplanes and, with 30 men, manufactured just as many Otto airplanes of his own design, including the first triplane (Oskar Wittenstein[2]) design) in 1911. Gustav Otto also constructed racing monoplanes for himself and the Munich physician Otto Erich Lindpaintner[3]. In addition, several types of monoplanes and biplanes with front-mounted engines were built, which was still a technical challenge at that time. On February 20, 1912, a world record with 5 passengers was set with such an Otto biplane (with a 100 hp 4-cylinder Argus engine in the front). But also the Otto airplanes with rear-mounted engine according to the Farman principle were quite successful. In the years up to 1912, they won about 100,000 marks in prizes and were also used by the Bavarian army administration.

When co-founder Herbert Alberti left the company in 1911, it was renamed Gustav Otto Flugmaschinenwerke. Gabriel Letsch, a designer from Alsace, headed the design department with great success in the years that followed

Above: Advertising AGO type 2 (above) and AGO type 3 (below).

(until the 1930s).

When the Royal Bavarian Flying Corps was founded in Schleißheim, Otto took advantage of his contacts with the military authorities and supplied the unit with his biplanes. When he presented his first military biplane in 1912, the Bavarian military administration placed an order for 40 aircraft. In order to obtain orders in Prussia as well, Otto founded the AGO-Flugzeugwerke in Berlin-Johannisthal in the same year. These remained standard aircraft for the young Royal Bavarian Air Force until 1914, but proved too weak

Above: Otto monoplane (#1) and biplane (#3)

Above: Otto-Flugmaschinenwerke on the Oberwiesenfeld near Munich (1911-1915). Hangar with an Otto mono- and biplane.

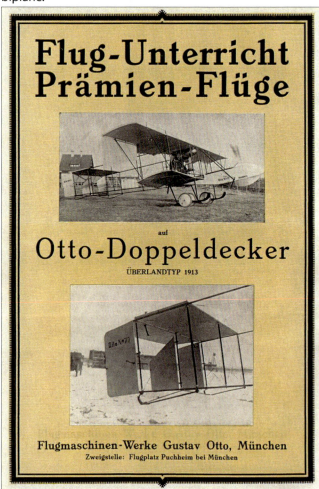

Above: Otto Biplane #77 of 1913. So-called premium flights are advertised as part of the national flight donation tender (Nationalflugspende).

for front-line use and were subsequently used for training purposes. His company in Munich therefore had to cease aircraft construction as early as 1915.

The factory was moved from Schleißheimer Strasse 135 to Oberwiesenfeld. Shortly after the beginning of World War I, the production of airplanes was accelerated. In 1914 a new Otto biplane with a front-mounted engine was built, for which a 150 hp Rapp engine was intended. This machine was the first to reach a speed of over 150 km/h and was described in the English trade journal "*Flight*" as a "noteworthy front-line opponent". In 1915, the so-called "Doppelrumpf-Kampfflugzeug" with a 200 hp Rapp engine (Ot C.I), an armed twin-fuselage biplane developed by the Swiss designer August Haefeli, was also built. It was used as a reconnaissance plane and was extremely popular with pilots because of its very good flight characteristics.

As early as 1912, Mr. Fremery took over the Otto-Werke agency founded in Berlin-Johannisthal. It was thanks to him that the name "Ago" was introduced as the official company name[4], recalling Gustav Otto's aircraft engine development (see also next page). A little later, the independent Ago-Flugzeugwerke, Berlin Johannisthal, emerged from this branch.

Gustav Otto himself was quite active. He was also a co-founder of the Pfalz-Flugzeugwerke, Speyer, in which he was

Above: Academy for aviation, founded by Gustav Otto as a flight school.

involved for a long time (see also Section 4.28).

At the end of 1915, a major disruption in engine deliveries occurred, causing delays in the delivery of front-line aircraft. This was compounded by a serious illness of Gustav Otto, who was unable to pursue his business to a sufficient degree. The factory in Munich-Oberwiesenfeld itself ran into financial difficulties, so that closure of the excellently equipped factory facility became inevitable. In the same year, the entire premises were sold to Bayerische Flugzeugwerke AG, a branch of Albatros-Werke, Berlin, from which "Bayer. Motorenwerke GmbH" (BMW) was created. The BMW signet, a Bavarian rhombus emblem stylized into a propeller, still refers to the company's beginnings in aircraft construction. Otto, who no longer held a stake in "Bayer. Flugzeugwerke AG", founded the "Otto-Werke GmbH" in Munich for the production of engine accessories.

Shortly before that, various experimental aircraft were still being constructed at the Otto factories, which were characterized in particular by considerably reduced air resistance. However, the forced sale prevented the successful completion of these developments.

Gustav Otto founded other companies, which will be described only briefly:

A new plant, the **Aktiengesellschaft Otto**, in Oschersleben, founded in 1916 built only aircraft parts for various manufacturers; it did not build anything of its own.

Above: Otto-Werke, Development of own aeroengines (A.G.O. - Aeromotor Gustav Otto).

After the end of the war in 1918, production was switched to automobiles, but this was not very successful, and the company was forced to close in 1928 as a result of the world economic crisis.

The **A.G.O. - Aeromotor Gustav Otto**: progress was also made in engine construction. Engine production was carried out by a specially created company, which also bore the abbreviation A.G.O. However, the letters with a period here stood for Aeromotor Gustav Otto.

The engineer Hans Geisenhof developed a number of promising light but powerful aircraft engines, but 100 hp Argus engines were also built under license at A.G.O.

The aircraft engines developed by Geisenhof were based on an interesting concept: the modular system. He developed a four-cylinder and a six-cylinder from each of two series.

It made absolute sense to have a separate company for engine construction, since from 1913 onwards the aircraft engines were procured separately by the air forces, tested and then, after testing, made available to the aircraft factories for installation. The use of alternative engines had also been considered by Otto in the design of his flying machines.

Above: Early Otto monoplane.

Above: The number on the fin identifies the aeroplane as the Otto-Alberti No. 2.

However, the Gnôme engines he also used were no longer allowed to be procured after 1912. French aircraft or engines were no longer desired by the German military. The engines used by the air forces came from Argus, Daimler, or the Rapp-Werke in Munich.

4.1.1 Otto-Werke GmbH, Flugzeug- und Maschinenfabrik GmbH, München (Ot)
Foundation:

Otto-Werke was founded on February 1, 1916 by Mr. Schrittisser and Gustav Otto. It emerged from the company Otto-Werke, Flugzeugwerke München (Ago, Gustav Otto), founded by Gustav Otto in 1909.

Factory Facility and Airfield:
The company had its own premises, which had a floor area of 1000 m² at the time of foundation but had increased to approx. 15,000 m² by the end of the war in 1918. The location proved to be favorable, as the production facilities were close to the Oberwiesenfeld airfield. This airfield was used for flying in and acceptance tests of the aircraft.

Fabrication:
The company itself was divided into
Division I Ignition bodies, wing gasoline pumps,
Division II F.T. couplings,
Division III Aircraft construction (new buildings and repairs, partly spare parts production).
New aircraft (manufactured under license) were of the Lvg B.I(Ot) type.
The performance of Otto-Werke developed as follows:

- a) From August 1914 to December 1914, three aircraft per week;
- b) From January 1915 to December 1915, five aircraft per week.

The output of Otto-Werke at its best times (end of 1915) was 45 aircraft per month, but mainly repairs were carried out. Department III (aircraft construction) also produced aircraft transport cars, of the following types:
- a) a lighter one for home planes to carry machines with light engines, and
- b) one for the front, on which aircraft with 200 to

Above: Otto-Werke: New assembly hall No. I in Oberwiesenfeld (100x40 m).

Above: Flight school hangar. In front on the right side is stored an early Otto biplane, in the back a LVG B.I(Ot).

260 hp engines could be loaded.
It can be expected that together with the prewar production around 350 aircraft were built at Otto in Munich.

AGO & Otto

Above: Maps showing the location of the Otto-Werke.

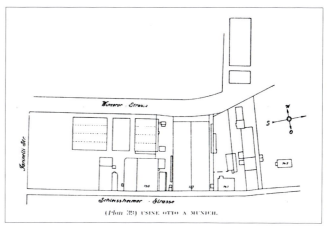

Above: Street map of Munich showing the Otto-werke.

Staff and Designers:
During World War 1, Otto-Werke employed up to 700 people[6] and built a total of around 110 aircraft by 1916.

Aircraft Development:
Since 1916, Gustav Otto, together with Mr. Direktor Schrittisser, had founded the new Otto-Werke GmbH at Schleißheimer Strasse 135 in Munich, initially continuing the previous operations. The share capital was originally 75,000 Marks, but was increased to 300,000 Marks at the beginning of 1918. Otto and Schrittisser continued to be the sole shareholders. The reason for the capital increase was the strong development of the company since the 2 years of its existence.

The equipment in the new workshops was extremely primitive. A machine park consisting of lathes and turrets, which had previously served for the production of primers of the old company, had to be repaired and renewed immediately. Tools and materials were lacking in all places, and the workrooms were gloomy.

The company's most urgent task was thus to acquire the new equipment necessary for production. This was all the more difficult as the economic situation in Germany began to deteriorate after 2 years of war. It took more than half a year until in October 1916 the performance of Otto-Werke was stable and delivery deadlines could be met.

In addition to the production of detonators and accessories for the emerging radio technology, aircraft construction was to develop as the third and strongest pillar of Otto-Werke GmbH. With 3 employees, spare parts for airplanes were produced on a smaller scale in July 1916, until the strong demand of the army administration encouraged the development of the factory here as well. Aircraft construction took up space, so adjacent land and buildings were purchased. In the immediate vicinity of the factory, a well-equipped carpenter's shop was acquired and expanded to meet the company's own requirements. For office and administration space, which had previously been 10 minutes away from the factory site, an intermediate building could be purchased, so that a more or less uniform factory site of about 15,000 m^2 could be created.

Until the end of the war, the workload of the aircraft construction division was intensive and high. In addition to the production of spare parts, new LVG B.I(Ot) machines were built, repairs were carried out and up to 45 machines were delivered per month.

Some special articles, like wooden wheels for airplanes, deck jacks for supporting and lifting airplanes, as well as airplane transport cars of own development completed the production range. The aircraft transport cars were built in 2 different versions, a lighter one for the training aircraft, i.e. to carry aircraft with lighter engines, and a heavier one for the front, on which aircraft with 200 and 260 hp engines could be loaded.

In spite of all the difficulties caused by the use of wooden

Development of Aircraft Factories

Aircraft Stocks in 1911					
Type	Engine [hp], Manufacturer	Crew	Armament	Year	Remarks
LVG B.I(Ot) LVG B.II(Ot)	100, Merc. D.I, 110-120, As, Bz, Merc.	2	– 1 MG	1915	64 examples, trainers
Otto C.I	200, Rapp 160, Merc. D.III 150, Benz Bz.III	2	1 MG	1915	18 expl. for Bavaria and Bulgaria, Prototype known as Otto KD15
Otto C.II	150, Benz Bz.III 160, Merc. D.III	2	1 MG	1915/16	24 expl.*, trainers

* 12 may have been completed with Alb C.III(Bay)

Left: Otto biplane on floats (1912).

Below: Second variant of Otto's own B-type. No series production took place, instead license production of the LVG B types.

Above: Early Otto B-type aircraft with Rapp engine.

Above: Otto C.I 545/15.

Right: Otto C.II aircraft with machine gun.

Above: Otto C.I (1915)..

Above: LVG B.II(Ot).

Left: LVG B.I(Ot) (1915)

Above: Otto C.II (1915/1916).

wheels with iron tires instead of the earlier rubber tires, this transport wagon has retained its favorable use and its absolute practicality unchanged. By means of a special superstructure, it was even possible to transport G-planes with their large profiles.

Production Supervisors (Bauaufsicht): #24

At Otto-Werke, there were two construction supervisors, one for aircraft production and a second for F.T. coupling construction (radio equipment).

Foundation:

Gustav Otto, supported by Director Fremery and his wife, founded AGO-Flugzeugwerke GmbH in Berlin-Johannisthal before closing his plant in Bavaria on May 1,

4.1.2 AGO-Flugzeugwerke GmbH, Berlin-Johannisthal (Ago)

1912, and at the same time supported the founding of Pfalz-Flugzeugwerke in Speyer. The reason for this step was a simple one: business with the Prussian Army Administration. Until then, business at the Munich site had been limited to the Royal Bavarian Air Force.

Within a very short time, Otto had established a monopoly position among the suppliers and outfitters of the Bavarian air forces. At first, they were of the opinion that one factory was sufficient to meet their demand for flying machines. Gustav Otto had told the Bavarian ministry that he could design and manufacture any type of aircraft it requested. Officials in Bavaria still refused to include Prussian aircraft manufacturers in their own procurement plans. Otto thus remained the monopolist. However, the head of the Bavarian Corps of Engineers criticized this state of affairs, recognizing that healthy competition could only be beneficial to the troops and that the companies, in constructive competition, would hardly ruin each other. Concerned for his company, he sought out new potential customers in Prussia, with Gustav Otto joining the throng of competitors in Berlin-Johannisthal, where he initially rented a few hangars at the "Neuer Startplatz" (New starting point).

The company took over a total of three sheds on the Johannisthal site in 1913, and production of Otto aircraft at the new site began as early as the beginning of 1913. In the years that followed, they developed into the strongest aircraft manufacturing company at the "Neuen Startplatz". Various versions of the "Otto trellis fuselage biplane" were initially built here, and a little later the plant was one of the most efficient Johannisthal armaments factories.

Factory Facility and Airfield:

The factory covered a terrain of about 48,000 m², was located at the Johannisthal airfield and had its own railway siding. The factory area developed as shown in the table below:

Development of Aircraft Factories

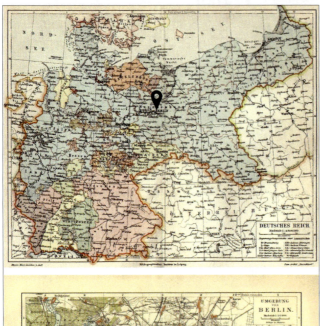

Above: Maps showing location of AGO-Johannisthal (Ago).

Above: Advertisement AGO Flugzeugwerke Berlin-Johannisthal (1913).

Fabrication:

The AGO company was engaged in the design and manufacture of amphibious aircraft, i.e. usable as land and seaplanes, with rear- or front-mounted engines. Ago-Flugzeugwerke designed a fighter aircraft, which was the first German fighter aircraft to go to the front under the type designation C.I (twin fuselage aircraft with rear-mounted engine and pressurized propeller arrangement). The company entered actual series production of warplanes in May 1915 and produced an improved type Ago C.II and Ago C.III at the end of 1915, although the latter did not get much beyond its test phase. Based on the experience gained, the Ago C.IV type was created, of which larger series were produced. In early 1918, the company delivered a larger number of Lvg C.II(Ago) biplanes. The following table gives an overview of the delivered aircraft:

With regard to seaplanes, the AGO pursued this with little interest. In 1915 and 1916, respectively, they developed two-seat floatplanes (150 hp Benz and 220 hp Benz engines) from the Ago C.I and Ago C.II landplanes. AGO-Flugzeugwerke in Johannisthal also created a type of giant aircraft, but the project was unsuccessful.

The flight school founded by Gustav Otto in Johannisthal was under the command of Lieutenant (ret.) von Gorissen, who had been trained by Eiler. Kießling, Winter and Breitbeil served as flight instructors.

In 1914, AGO Flugzeugwerke built the first major maintenance for seaplanes at Priwall near Travemünde.

Staff and Designers:

For the period since the foundation of the company and the outbreak of the war there is no information about the staff.

Time	Middle 1914	Spring 1915	Spring 1917	Spring 1918
Floor Aera	1,500 m²	5,500 m²	10,500 m²	27,000 m²

AGO & Otto

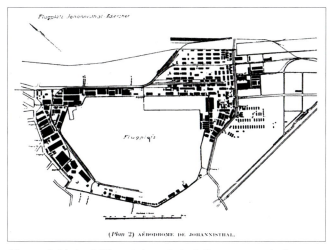
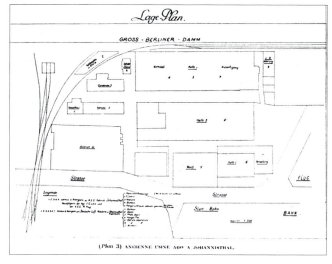

Above & Right AGO at Johannisthal, Situation end of 1918.

Above: AGO hangars in Johannisthal (1914).
Left: Advertisment AGO Johannisthal: Landplanes, Seaplane, Flight school (1913).

Besides the technical director Fremery, Mr. Gabriel Letsch acted as the chief designer of the company.

Aircraft Development:
Starting in 1913, the Johannisthal-based company AGO-Flugzeugwerke GmbH was engaged in the design of experimental aircraft and the construction of flying machines. No specific details can be given about the work that was started in many different ways and in some cases not completed before May 1915.

In May 1915, AGO delivered an aircraft, a fighter, under the type designation "Ago C.I". This type was soon followed by a similar "C. II" and then, after its beginnings, the "Ago C III", another fighter aircraft with the type designation "Ago IV". This type was probably the most successful of the company, which is best illustrated by the following delivery table. At the beginning of 1918, the company built a larger number of "Lvg C.II(Ago)" aircraft under license and, in parallel with these deliveries, was engaged in the design and manufacture of a C aircraft Ago C VI and C VII.

Production Supervisors (Bauaufsicht): #3
Production Supervision No. 3 consisted of 1 officer, 1 representative (engineer) of the Inspectorate of Airborne Troops, 1 assistant sergeant, 1 vice sergeant, 2 privates, and 4 men.

War Material Found by The Inter-Allied Commission of Control (Iacc) During Inspections After the End of the War:
2 aircraft Ago S.I
19 aircraft Lvg C.VII (16 without engines)

Aircraft Delivered by Ago					
Year	1914	1915	1916	1917	1918
Number		32	62	233	76
Type		C.I & C.II	C.I, C.II, & C.IV	C.IV	LVG C.II, C.VIII

Development of Aircraft Factories

Above: Naval Ago C.II.

17 different aircraft engines
94 semi finished fuselages of Lvg aircrafts

Situation after WWI:
Since 1919, the factory had ceased all work in the field of aviation. In 1920, it employed 760 workers in the production and repair of all types of railroad cars in large series.

After the armistice, the "Ago" factory in Johannisthal began manufacturing doors and windows. On June 1, it became the property of AEG, and the production of doors and windows continued until January 1, 1020.

The factory was closed, two large halls were rented to Maschinen-Fabrik Oberschönewalde A. G, another was rented to Maschinen Verkaufs Genossenschaft and a

Below: Ago C.I (1915).

Above: Ago C.III (1915).

Above: Ago DV3: single-seat biplane with 100 hp Oberursel U.I rotary engine, normal design and sweep, fully cowled engine, built 1914/15. A two-seat versionis also known,

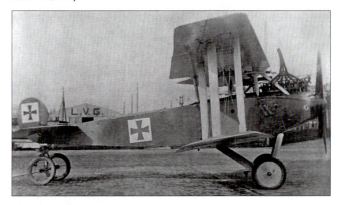

Above: Ago C.IV.

Above: Ago C.II (1915).

Above: Ago C.IV (1916).

AGO & Otto

Ago Aircraft					
Type	Engine [hp], Manufacturer	Crew	Armament	Year	Notes
Landplanes					
Ago C.I (DH 6)	150, Benz Bz.III 160, Merc. D.III 160, Mayb. Mb.III	2	1 MG	1915	Double fusealge
Ago C.II (DH 7)	160, Merc. D.III	2	1 MG	1915	Double fuselage, 3 examples at front line
Ago C.III	160, Merc. D.III	2	1 MG	1915	No series production
Ago C.IV	200, Benz Bz.IV	2	2 MG	1916	~150, Schül: 66, Rat*: 10
Ago C.V	160, Merc. D.III	2		1916	No series production
Ago C.VI	200, Benz Bz.IV	2		1916	No series production
Ago C.VII	200, Benz Bz.IV	2		1916	1 prototype
Ago C.VIII	260, Merc. D.IVa	2		1917	1 prototype
Ago C.IX (CL.VIII)	260, Merc. D.IVa	2		1917	1 prototype
LVG C.II(Ago)	160, Merc. D.III	2		1917/8	300
Ago S.I	300 oder 500, Bus oder Benz	2	2 MG, 1 Cannon	1918	2 prototypes
Seaplanes					
Seaplane Trainer	100, Argus As.I	2		1912	Marine D7 (pusher)
Seaplane Trainer (W-1)	150, ArgusAs.III	2		1914	Marine D15
D.19 Floatplane Trainer (copy Avro 503)	100, Gnome	2		1913	Marine 30 - 39, 112, 113
Marine DD 1914 (Land)		2		1914	Marine S.16
Ago Seaplane Biplane	150, ArgusAs.III	2	2 MG		Marine 65 – 69 (tractor)
Ago DHW 2	150, ArgusAs.III 200, Oberursel U.II	2	2 MG		70, 71 72, 114
Ago C.Iw (See)	175, Benz Bz.III	2	2 MG	1915	115
Ago C.IIw (See)	175, Benz Bz.III 200, Benz Bz.IV 235, Merc. D.IV	2	2 MG	1916	539, 586

* Waggonfabrik Jos. Rathgeber, München-Moosach.
Mention should also be made at this point of the Rathgeber company, which was given the abbreviated designation "Rat" by the Flugzeugmeisterei. This company was also located in Munich (near the Munich-Moosach railroad junction) and, like Otto-Werke, was assigned to Building Inspectorate No. 24 (see further information on the Building Inspectorates in a separate chapter). In 1916, the company began building aircraft and initially received an order for the construction of 10 new Ago C IV(Rat) aircraft. Due to a series of difficulties, mainly caused by the licensing company Ago (Johannisthal), and the impossibility to provide itself with suitable personnel, the company exceeded the usual delivery deadlines to such an extent that a further order for new construction was not placed and the company was only called upon to carry out repairs. After the delivery of the first 10 repair aircraft, Idflieg renounced Rathgeber also as a repair company, so that after the delivery of the Prussian orders the company only repaired for the Bavarian Inspectorate of Military Aviation. Monthly performance averaged 20 repair aircraft. According to the report of the Interallied Aviation Inspection Commission (ILÜK), a total of about 300 aircraft were repaired.

Above: Ago C.V (1916).

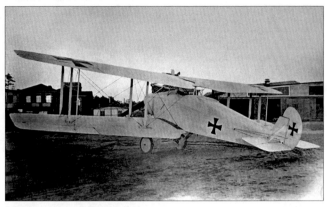
Above: Ago C.VI (1916).

Above: Ago C.VII (1917).

Above: Ago C.VIII (1917).

Above: Ago C.IX (1917).

Above: Ago WDD (1913).

Above: Ago C.Iw (1915).

Above: Ago C.IIw (1916).

Above: Ago E.I (1913)

Above: Ago S.I (1918).

fourth to Deutsche Luft Reederei, which used the hall as a workshop for its civil aircraft.

Endnotes

1 See Scheer, AGO-Flugzeugwerke.
2 Oscar Jürgen Wittenstein, also Oskar Jürgen Wittenstein (* September 28, 1880 in Barmen; † September 3, 1918 in Rudow near Berlin) was a German industrialist, art collector and aviation pioneer. Wittenstein was trained as a pilot in 1909 by French aviation pioneers Henri and Maurice Farman. He also took a pilot's examination on a Farman biplane at the Puchheim airfield in the Fürstenfeldbruck district on March 18, 1911. His "German pilot's license" was issued in Munich on April 29, 1911, and bears the number 81. He was the first person to fly over the city of Munich, and he became financially involved in the first airship club. In 1911, together with Adolf Erbslöh, Wittenstein founded Flugwerk Deutschland in Munich and Luftschiffbau-Gesellschaft Veeh mbH, so named after the inventor Albert Paul Veeh from Apolda. In 1910/1911, a triplane with staggered wings was built, but no flights are known. Also around 1911, a monoplane with a 50-horsepower Argus engine, two elevators and an uncovered tubular steel fuselage was manufactured and is said to have flown in 1911. For financial reasons, they soon had to abandon the company. Wittenstein took part in numerous flying competitions, including with his wife's brother Hans Robert Vollmöller. In 1911, for example, he took part in the first competition of the Deutschlandflug. In 1916, Wittenstein was involved in the design of the giant AEG R.I airplane in Munich and was considered an employee of the company. The aircraft, equipped with 4 × 260 hp Daimler engines centrally in the fuselage in two rows side by side and a wingspan of 36 meters at 19 length, crashed on September 3, 1918 due to engine failure, killing seven people, including the young pilot Lieutenant Brückmann as pilot, the engineer Otto Reichardt (1885-1918) and Oskar Wittenstein.
3 Otto Erich Lindpaintner (auch: Otto Erik Lindpaintner; * 2. März 1885 in München; † 22. Juli 1976 ebenda) war ein deutscher Arzt, Flugpionier und Rennfahrer. Zwischen 1910 und 1914 errang er auf verschiedenen eigenen Flugapparaten zahlreiche Preise.
4 Nun: „Aktiengesellschaft Gustav Otto"
5 50,000 HZ 05 ignition bodies were delivered for the Royal Main Laboratory in Ingolstadt. This was followed by smaller orders for gasoline pumps and accessories for F.T. units (radio systems).
6 According to the IACC report from 1920, the workforce of Otto-Werke in Munich grew from 15 employees, at the beginning of 1916, to 1,100 at the end of 1917.
7 When Gustav Otto established his Ago subsidiary in Johannisthal, four competing companies were located at the so-called "Neuer Startplatz" alone: Rumpler Luftfahrzeugbau GmbH, Harlan-Flugzeugwerke GmbH, Automobil & Aviatik AG and A.H.G. Fokker Aeroplanbau. The "old starting place" was also home to another important manufacturer: Albatros-Flugzeugwerke GmbH.

Above: Ago C.V (1916).

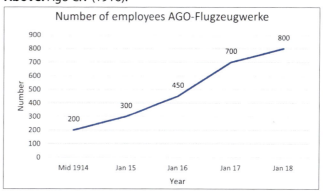

Development of Aircraft Factories

4.2 Albatros-Gesellschaft für Flugzeugunternehmungen m.b.H., Berlin-Johannisthal (Alb)

4.2.1 Albatros-Werke GmbH, Berlin-Johannisthal (Alb)

Foundation:
Captain (ret.) Dr. Walther Huth[1] carefully studied the development of aviation, especially the progress in France.

Believing in the positive signals and in the conviction of a profitable business, Dr. Huth had a hangar built for 6,000 Marks at the first German airfield at Johannisthal near Berlin. The company founded here under the name "Ikaros-Gesellschaft" intended to carry out demonstration flights all over Germany with one and two-seater Antoinette monoplanes. In 1910, a Farman III biplane was added. At the first international air show in August 1909, the Grande Semaine d'Aviation de la Champagne, the Farman machine had won the long distance prize and the Antoinette the high altitude prize. These machines were used as training aircraft and as patterns for license replication in the later Albatros factories. In addition to the Antoinette and Farman types, the three best-known Albatros replicas were the French

Above: As one of the very first German aircraft factories, this plant was founded on December 20, 1909 by Dr. Walter Huth under the name "Pilot" - Flugtechnische Gesellschaft".

summer biplane. All aircraft were successfully flown and used in student pilot training.

In the absence of trained pilots, Dr. Huth sent his chauffeur Simon Brennhuber[2] and the employed engineer Eugen Wiencziers[3] to France to learn to fly.

On December 20, 1909, the „Ikaros-Gesellschaft" was renamed „Pilot-Flugtechnische Gesellschaft". Initially, no aircraft were designed, built or sold. The French airplanes, which had arrived in the meantime, were demonstrated at various air shows, mainly by Wienczier, and won several prizes.

On December 29, 1909, Dr. Huth renamed the company "Albatros-Werke" to implement his acquired license rights to build French airplanes. As early as the spring of 1910, with the participation of other financiers, including engineer

Above: Advertisement 1914, *Motor*.

Albatros

Above: View of the Albatros production hall.

Above: Maps showing location of Albatros Werke.

Wars of 1912-1913. One of them flew its first military mission in the skies over Europe on October 16, 1912.

After Ernst Heinkel, who served as chief designer from 1912 to 1914, was hired, significant in-house designs in the form of two-seat bombers and reconnaissance aircraft followed from 1913, while equally significant single-seat fighters were built under Robert Thelen.

On March 2, 1917, "Albatros-Werke GmbH" Otto Wiener[4], the "Albatros-Werke" was transformed into a limited liability company GmbH).

The Farman aircraft, imported from France by Albatroswerke, was sold to the Army Administration on December 18, 1910, having already been loaned out for military pilot training at Döberitz. Under the Army designation "B.I", it was the first German military aircraft. Among the five other "flying machines" acquired by the German Army Administration at the end of 1910 were a Farman and a Sommer[5] biplane, two French types copied from Albatros. In 1911, further important development steps followed with the first in-house design, the MZ 1 twin pigeon, and with the victory of Benno König[6] on Albatros-Farman in the Deutschlandflug.

During 1912, five Albatros F.2s were built, a further development of the French Farman III biplane (hence the letter F) with a gondola for the crew and Argus in-line engine instead of the original Gnôme rotary engine. Four aircraft were delivered to the newly formed Bulgarian Aircraft Department, where they participated in the Balkan

Above: Albatros advertisement 1914.

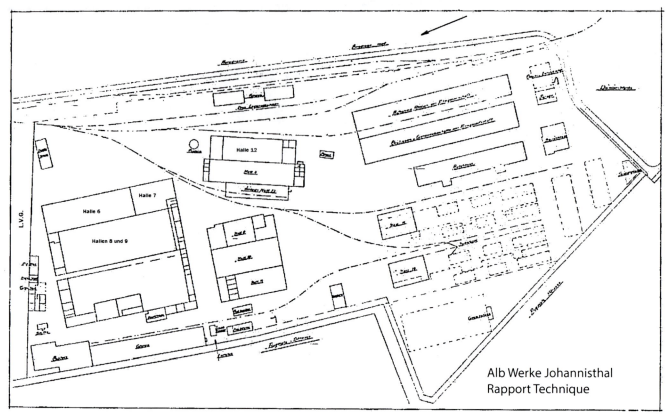

Albatroswerke in Johannisthal

liquidated and the company "Albatros-Gesellschaft für Flugzeugunternehmungen mbH, Berlin-Johannisthal" emerged from the liquidation. The company continued to operate under Dr. Walter Huth and Otto Wiener.

Albatros became known for building some of the most famous and best fighter aircraft of World War I for the air forces of the Empire, such as the Albatros D.III or the Albatros D.V. At the beginning of the war, the number of employees was 830, and by November 1918, it was already over 5,000. By the end of 1918, around 10,400 aircraft had been built.

The branches of the "Albatros-Gesellschaft für Flugzeugunternehmungen mbH" were:

1. in Friedrichshagen near Berlin: „Albatros-Gesellschaft für Flugzeugunternehmungen mbH, Berlin -Johannisthal, Abteilung Wasserflugzeugbau und Reparaturwerkstatt Friedrichshagen" (founded September 1, 1916). (see chapter 4.2.2);
2. in Schneidemühl (Pomerania): „Albatros-Gesellschaft für Flugzeugunternehmungen mbH, branch Ostdeutsche Albatros-Werke GmbH, which had been given the abbreviated designation „Oaw" by the Flugzeugmeisterei. (see chapter 4.2.3);
3. in Warsaw: „Albatros-Gesellschaft für Flugzeugunternehmungen mbH, Berlin -Johannisthal, Abteilung Militärwerkstätten Warschau", founded on January 1, 1916, closed on April 1, 1918. The „Albatros-Gesellschaft für Flugzeugunternehmungen mbH" in Warsaw, received the short designation „Refla" from the Flugzeugmeisterei. (see also 4.2.4).

Albatros Werke Factory and Airfield at Johannisthal:

In 1909, operations were started in a hangar of about 1,440 m² located on the south side of the Johannisthal airfield.

In 1910, production is moved to its own factory building, which has been built in the meantime. Total area is now already 2,040 m².

During the war, there was a support fund from which those women whose husbands had previously been employed at the plant but had been deprived of their family through military service were supported. These women also received special allowances for their children at Christmas.

A factory canteen provided lunch for the workers employed at the plant.

For the lighting and power systems, electricity was obtained from the municipal power plant and transformed to the required voltage of 220 volts. The majority of the machines had separate supply lines, so that in the event of malfunctions the entire machine stock was not affected.

After the outbreak of World War I, war planes were built, and the factory facilities were expanded accordingly for this purpose. In September 1916, a small plant was added at Friedrichshagen near Berlin, where mainly seaplanes were built. At the instigation of the ldflieg, the military repair plant "Refla" had been established in Warsaw in the same

Albatros

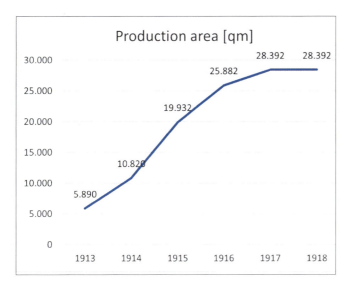

Left & Above: Growth of Albatros Werke Johannisthal.

year to provide a repair facility to relieve railroad traffic near the eastern front. Later, as the number of machines to be repaired had declined sharply, and in order to keep the workforce fully occupied, a small number of B.IIs were rebuilt.

During World War I, Albatros built a total of about 10,400 aircraft, of which 6,390 were built in Johannisthal, about 306 in Friedrichshagen - including 150 seaplanes - and about 3,650 in Schneidemühl at OAW. Albatros aircraft were also reproduced at the Austrian "Phönix-Werk" near Vienna.

Fabrication:
The "Albatros" factory in Johannisthal already existed before the war.

During the war, it achieved an annual maximum production rate of 2,300 aircraft and manufactured a total of 3,012 aircraft of various types.

Staff and Designers:
The company's design engineers were: Engineer Gabriel, Engineer Heinkel, Dipl.-Ing. Thelen (Technical Director), Dipl.-Ing. Schubert (Design Manager), Senior Engineer Hüttner (Technical Director).

The staff grew permanently, as can be seen from the adjacent diagram.

Former pilots of the company and their special achievements:

König: Winner in the German round flight 1911 and winner of the prize of the Ministry of War

Hirth, Hellmut: Winner in the Lake Constance Seaplane Competition 1913, winner of the Grand Prize; Winner in the International Seaplane Competition on the Upper Italian Lakes, Fall 1913.

Vollmöller: winner of the honorary prize in the Lake Constance Water Flight 1913.

Kühne: 2nd overall prize winner in the East Prussian Round Flight.

Landmann: set an endurance world record of 24 hours flying time on June 27, 1914.

Dipl.-Ing. Thelen: Honorary prize of the Grand Duke of Baden at the Lake Constance Competition in 1913; Achieved 800 m in 7 minutes on Albatros biplane with military load at the Autumn Flight Week in 1913 and received the War Ministry's prize for this; He flew the Albatros biplane with passenger for the National Flying Award and covered 1,400 km in one day; Climbed 2,850 m on Albatros biplane with four passengers and set a new record.

Aircraft Development:
1909/10: The first design launched by Albatroswerke was a monoplane based on the French Antoinette type. A 50 hp Antoinette engine was installed in the 1909 model on an experimental basis, and a 100 hp Gnome rotary engine in the 1910 model.

1911: In this year the company released an improved Sommer-Farman type, equipped with 70 hp Gnome 7-cylinder rotary engine. The flight requirements placed on the flying machine were: Minimum speed: 60 km/h. The aircraft must be able to fly for 30 minutes at an altitude of at least 300 m with full load. The takeoff distance was not allowed to exceed 50 m.

At the same time, a military type was also created: M.Z. 2, a biplane with a 100 hp Argus engine weighing 480 kg, equipped with two protected seats and dual controls with levers. The engine was located in the rear, pressurized propeller. Speed was 80 km/h. At the end of the year, another Farman type with 50 hp Gnome and 70 hp Renault engine was developed. The required performances were now: Minimum speed 60 km/h, 2 hours flight time at minimum altitude 300 m.

1912/13: During this period the military type 1912 (VM.Z. 1) was built, which was a biplane equipped with a 100 hp Argus engine. The total weight was 420 kg, the payload 300 kg. Fuel was available for 4 hours, maximum speed was 90 km/h.

Albatros Aircraft

Type	Engine [hp], Manufacturer	Crew	Armament	License Built By	Year	Notes
Landplanes						
Alb A types	diverse	2	–		1914-15	>50, A, A.I, A.II
Alb B.I (L1, DD.1)	100, Merc. D.I	2	–	OAW Refla	1914-15	>200
Alb B.II (L2, DD.2)	100, Merc. D.I 110, Benz Bz.II 120, Argus As.II	2	–	Av: 200 Rol: ~830 Mer: >400 Bay: ~180 Refla: 100 Kon: 350 Li: 150	1914–1915	Total orders 1914-18: 3,544
Alb B.IIa (L30)	100, Merc. D.I 110, Benz Bz.II 120, Argus As.II	2	–		1917-18	
Alb B.III (L5)	100, Merc. D.I 120, Merc. D.II 150 Benz Bz.III	2	–	OAW: ~150	1915	Trainer
Alb C.I (L6)	150, Benz Bz.III 160, Merc. D.III	2	1 MG	Rol: 79 (aka Rol C.I)	1915–17	~400
Alb C.Ia (L32), **Ib** (L33), **Ic, Id, If**	185, Argus As.III 160, Merc. D.III	2	1 MG	Bay: 370 Mer: 500 Rin: 150	1917	
Alb C.II (L8)	150, Benz Bz.III	2	1 MG		1916	1 prototype, pusher
Alb C.III (L10)	150, Benz Bz.III 160, Merc. D.III	2	2 MG	Bay (Bay C.I): 655 Li: 75 Hansa: 200 LVG: 300 DFG: 200 OAW: 300 SSW: 93	1915-18	468
Alb C.IV (L12)	160, Merc. D.III	2	1 MG		1916–17	24
Alb C.V (L14)	220, Merc. D.IV	2	2 MG	OAW: 25	1916–18	100 + 3
Alb C.VI (L16)	180, Argus As.III	2	2 MG		1916–17	>75
Alb C.VII (L18)	200, Benz Bz.IV	2	2 MG	Bay: 275	1916-18	225
Alb. C.VIIIN (L19)	160, Merc. D.III	2	2 MG, bombs		1917	5
Alb C.IX (L23)	160, Merc. D.III	2	2 MG		1916-18	3 prototypes
Alb C.X (L25)	260, Merc. D.IVa	2	2 MG	Rol: 100 Li: 50 Bay: 50 OAW: 200	1916–18	
Alb C.XI (L26)						prototype
Alb C.XII (L27)	260, Merc. D.IVa	2	2 MG	Li: 0 (25), canc. Bay: 180 OAW: 150	1917–18	150 + 3
Alb C.XIII (L29)	160, Merc. D.III	2	2 MG		(1917)	prototype
Alb C.XIV (L31)	220, Benz Bz.IVa	2			(1918)	prototype
Alb C.XV (L47)	220, Benz Bz.IVa	2	2 MG		1918	>10

Albatros

Albatros Aircraft						
Type	Engine [hp], Manufacturer	Crew	Armament	License Built By	Year	Notes
Alb D.I (L15)	160, Merc. D.III	1	2 MG		1916-18	50
Alb D.II (L17)	160, Merc. D.III	1	2 MG	LVG: 75	1916-18	200
Alb D.III (L20)	160, Merc. D.III	1	2 MG	OAW: 840	1916-18	500
Alb D.IV (L22)	160, Merc. D.III	1	2 MG		(1917)	1 (3) prototype
Alb D.V (L24)	160, Merc. D.III	1	2 MG		1917-18	900
Alb D.Va (L24)	170, Merc. D.IIIa	1	2 MG	OAW: 600	1917-18	1062
D.VI (L28)	160, Merc. D.III	1	2 MG		-, (1918)	Prototype, pusher
D.VII (L34)	195, Benz Bz.IIIbo	1	2 MG		-, (1918)	prototypes
D.VIII (L35)	200, Adler Ad.IV	1	2 MG		-, (1918)	Not completed
D.IX (L37)	170, Merc. D.IIIa	1	2 MG		-, (1918)	prototype
D.X (L38)	195, Benz Bz.IIIbm	1	2 MG		-, (1918)	prototype
D.XI (L41)	160, SieHa Sh.III	1	2 MG		-, (1918)	prototypes
D.XII (L43)	170, Merc. D.IIIa 185, BMW BMW.IIIa	1	2 MG		-, (1918)	prototypes
D.XIII (L44)	170, Merc. D.IIIa	1	2 MG		-, (1918)	prototype
D.XIV (L46)	185, BMW BMW.IIIa	1	2 MG		(1918)	prototypes
Dr.I (L36)	160, Merc. D.III	1	2 MG		(1916)	prototype
Dr.II (L39)	195, Benz Bz.IIIbm	1	2 MG		(1917)	prototype
J.I (L40)	200, Benz Bz.IV	2	2 MG 1 LMG		1917–18	125
J.II (L42)	220, Benz Bz.IVa	2	2 MG 1 LMG		1918	150
Fok D.VII(Alb)	180, Merc. D.IIIa	1	2 MG		1918	~1000
Alb G.I (L4)	4x120, Merc. D.II	3			(1917)	1, built at OAW
Alb G.II (L11)	2x150, Benz Bz.III	3	2 MG, bombs		1917	~15
Alb G.III (L21)	2x220, Benz Bz.IV	3	2 MG, 300 kg bombs		1917	~5
Seaplanes						
WRE	75/100, Merc.	2	-		1913	1 (2 variants)
WDE	100, Merc. D.I 100, Argus As.I 75, NAG	2	-		1913	3 (based on Alb B.I)
W.1	150, Benz Bz.III, 160, Merc. D.III	2	1 Torpedo		1916	19
W.2	150, Benz Bz.III	2	1 MG		1916	1
W.3	2x150, Benz Bz.III	2	1 MG		1916	1
W.4	160, Merc. D.III	1	1 or 2 MG		1916/17	118
W.5	2x150, Benz Bz.III	2	2 MG + 1 Torpedo		1918	5
W.8	195, Benz Bz.IIIbo / bm	2	2–3 MG		1918	3 (8), 5001-5003

Above: Albatros Taube.

Above: Albatros biplane Taube.

Above: In terms of orders, the D.V was the most successful Albatros design. Earlier Albatros fighters were much more successful in combat; they were more robust and faced less advanced Allied fighters. The headrests were soon removed.

Above: The Albatros G.II was built in small numbers.

Above: A development of the G.II, the Albatros G.III had more powerful engines, simplifed landing gear, and simpled wing bracing for reduced weight and drag.

Albatros Two-Seater Specifications

Type	Max Speed [km/h]	Climb Rate [minutes]				
		1000 m	2000 m	3000 m	4000 m	5000 m
Alb C.I	132	9¾	25	58½		
Alb C.II			25			
Alb C.III	132	7½	18½	38½		
Alb C.IV		8½	20	42		
Alb C.V	170	4½	9½	16	25	
Alb C.VI		7	17	34		
Alb C.VII		6½	13	21	34	
Alb C.VIII		5	14	27	50	
Alb C.IX		4	9½	17	29½	58
Alb C.X		3	6½	11	21	49
Alb C.XI		Project only, not built				
Alb C.XII		3½	8	13	21½	33
Alb C.XV		2½	6¼	11¼	16¾	24
Alb C.XVa		3½	7¾	12¾	19¾	30¼

Albatros Fighter Specifications

Type	Max Speed [km/h]	Climb Rate [minutes]				
		1000 m	2000 m	3000 m	4000 m	5000 m
Alb D.I	175	4	9	16½	26	35
Alb D.II	175	3	5½	9½	13½	19½
Alb D.III	165	3½	7	12	1 ¾	29
Alb D.IV		4	8	13½	21	32
Alb D.V	165	1½	5½	10	16	27
Alb D.XI	190	1¾	4¼	6¼	9¼	12¾
Alb D.XII	180	2				26
Alb D.III	185	2				23
Alb D.XIV	185	2				23

Albatros Seaplane Specifications

Type	Max Speed [km/h]	Climb Rate [minutes]				
		1000 m	2000 m	3000 m	4000 m	5000 m
Alb W.1			12–15	40		
Alb W.2		4½	8	11	15	19½
Alb W.3		12	25	37	50	
Alb W.4	160	2	3¼	4¼	6	6½
Alb W.5		5½	10	14	20	

Development of Aircraft Factories

Albatros Bomber Specifications						
Type	Max Speed [km/h]	Climb Rate [minutes]				
		1000 m	2000 m	3000 m	4000 m	5000 m
Alb G.II		9½	25	70		
Alb G.III		10	30	45		

Another design was a seaplane for military purposes under the type designation W.M.Z. (Wasser-Militär-Zweidecker), which had the military designation D.5, designed as a biplane with a 100 hp Argus engine and a pressurized propeller. Acceptance of the aircraft took place in August 1912.

In further development, after the Bréguet type, a biplane with a 70 hp Renault engine was released, which had an hourly speed of 85 km/h with a 200 kg payload. The 1912 type, which preceded this biplane, was a three-seater with a 100 hp Argus engine that reached 500 m altitude in 10 minutes with a payload of 300 kg. The flight requirements for the same were: Action radius 250 km, climbing ability 500 m in 15 minutes, measured in circular flight.

To meet the further requirements, the company launched the D.E. type. It was equipped with a 100 hp Argus engine. The load capacity was also 300 kg. Fuel for 4 hours could be accommodated. The requirements were: Action radius 250 km, climbing ability 500 m in 15 minutes, minimum speed 80 km/h, were achieved or exceeded in the hourly speed, which was determined to be 90 km/h.

1913: In this year the company launched a monoplane (Taube) under the designation Type E.E. The climb time for 800 m altitude with military load (200 kg and 4 hours of operating fuel) was 10 to 15 minutes. The maximum speed was between 100 and 105 km/h.

1914: Further development of the Taube resulted in the type F.T., a military monoplane whose fuselage was made of plywood. The climb time was 800 m in 8 minutes with a prescribed payload (200 kg and fuel for 4 hours).

As a result of the outbreak of war, aircraft types were created to meet the requirements of the army administration.

The first type to emerge was the Alb B.I, a three-strut military biplane equipped with the 100 hp six-cylinder Daimler D.I engine. The aircraft still had side coolers and a 6-hour fuel tank. The climb time to 800 m was 8 ½ minutes.

Shortly thereafter followed the improved model with military designation B.II also with a speed of up to 108 km/h. The climb time to 8,000 m was reduced to 5 minutes with a full load (operating fuel for 4 hours with a payload of 200 kg).

The Albatros B.IIIs had a basically similar superstructure to their predecessors, but already achieved speeds of up to 140 km/h. The engines used were 100 or 120 hp Mercedes D.I and D.II engines or 150 hp Benz Bz.III. Climbing speeds continued to decline (at full load: 800 m in 12 minutes, 2,000 m in 40 minutes).

The C types that followed were so-called fighter two-seaters. Flight performance improved from type to type as as shown above and on the previous page:

Production Supervisors (Bauaufsicht): #1

The Bauaufsicht was established at the end of 1916. The leader of the construction supervision was an officer, to whom another 5 officers and 1 authorized engineer of the Inspectorate of Aviation Troops were assigned. In addition to the aforementioned, the staff of the Production Supervisors consisted of nine foremen, 4 non-commissioned officers, 13 recruits, and 4 test pilots.

War Material Found by the Inter-Allied Commission of Control (IACC) During Inspections after the End of the War:

192 licensed built Fokker D.VII(Alb) and different Albatros aircraft,
117 aircraft engines (Benz and Mercedes),
143 matrices and stamping tools specially designed for the production of aviation equipment.

Situation after WWI:

After the armistice and in 1919, it completed 100 Fokker D.VII aircraft.
The "Albatros" factory in Johannisthal was still spending heavily in 1919 to build new reinforced concrete workshops. In 1920 it did not work for the aviation industry. It manufactured wooden beds, patented slatted frames and centrifugal pumps.
In January 1920, it employed 500 workers; in November of the same year, it employed only 50.

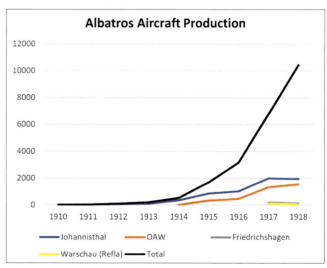

Albatros

Above: Alb A type (EE) (1913)

Above: Albatros Doppeltaube B.22a/12 (1913)

Above: Alb B.I (1914).

Above: Alb B.II (1914).

Above: Alb B.Ia (Argus) (1917).

Above: Alb B.IIa (Mercedes) on skiis (1917).

Above: Alb B.III (1915) (LF 171, Navy land plane).

Above: Alb C.I (1915).

Development of Aircraft Factories

Above: Alb C.Ia (1917).

Above: Alb C.If (1917) (Argu.s As.III).

Above: Alb C.II (1916).

Above: Alb C.III (1915).

Above: Alb C.IV (1915). (Thick airfoil testbed.)

Above: Alb C.V (1916).

Above: Alb C.VI (C.III airframe with Argus As.III)(1916).

Above: Alb C.VII (1916).

Albatros

Above: Alb C.VIIIN (1916).

Above: Alb C.IX (1916).

Above: Alb C.X (1917).

Above: Alb C.XII (1917).

Above: Alb C.XIII (1917).

Above: Alb C.XV (1918).

Above: Alb D.I (1916).

Above: Alb D.II (1916).

Development of Aircraft Factories

Above: Alb D.III (1916).

Above: Alb D.IV (1917).

Above: Alb D.V (1917).

Above: Alb D.Va (1917).

Above: Alb D.VII (1918).

Above: Alb D.IX (1918).

Above: Alb D.X (1918).

Above: Alb D.XI (1918).

Albatros

Above: Alb D.XII (1918).

Above: Alb Dr.I (1916).

Above: Alb Dr.II (1916).

Above: Alb G.III (1917).

Above: Alb J.I (1917).

Above: Alb J.II (1918).

Above: Alb W.1 (WDD) (1916).

Above: Alb W.2 (1916).

Development of Aircraft Factories

Above: Alb J.II (1918).

Above: Alb J.II (1918).

Above: Alb J.II (1918).

Above: Alb J.II (1918).

Above: Alb WRE (1913).

Above: Fok D.VII(Alb).

Above: Alb. C.V/16; ear radiators, square lower wingtips.

Above: Alb. C.V/17; airfoil radiator, rounded lower wingtips.

Albatros

4.2.2 Die Albatros-Werke Friedrichshagen bei Berlin

Unlike the branch plants in Schneidemühl and Warsaw, this plant was not given a separate military short designation.

Factory Facility:
The factory was located between a road and the Spree River in a location that was ideal for a factory producing seaplanes.

Fabrication:
The Albatros factory in Friedrichshagen mainly carried out repair work on aircraft. In addition, it produced a total of: both 300 land and 150 seaplanes and achieved a monthly production of 30 aircraft.

The planes built here were of the same types as those from the Albatros factory in Johannisthal.

Staff and Designers:
The maximum number of workers and civil servants was 576, with 500 engaged in aeronautical work during the war.

Aircraft development:
No aircraft were developed at the Albatros plant in Friedrichshagen. All development work was done at the main plant in Johannisthal.

War Material Found by the Inter-Allied Commission of Control (IACC) During Inspections after the End of the War:
At the end of the war there were still at this location: 23 Albatros aircraft in poor condition: 19 without engines and four with engines, and one W8 seaplane with engine.

Situation after WWI:
After the First World War, Albatros continued its work at the headquarters, initially under chief designer Rudolf Schubert, and from 1926 under Walter Blume, even after the end of the war. In 1924, for example, the L 59 single-seat sports aircraft was built, a wooden low-wing monoplane with a Siemens & Halske Sh 4 five-cylinder radial engine and 45 kW power. Only one example was built, which was followed in the same year by the two-seater L 60 with the Sh 5 (62 kW) and L 66 or L 66a. In 1925, the light aircraft L 67 and the school biplane L 68 based on the B.II were designed and built by the only 150 employees left. Later, only the L 82 and L 101 were added in significant numbers. In September 1931, the company was forcibly merged with Focke-Wulf-Flugzeugbau AG, which was subsequently called Focke-Wulf-Albatros for a short time.

Above: Maps showing the location of the Albatros factory in Friedrichshagen.

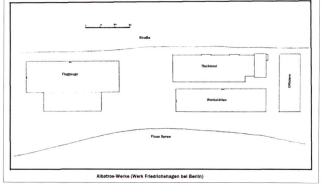

Above: Alb Werk Friedrichshagen facilities.

4.2.3 Ostdeutsche Albatroswerke GmbH, Schneidemühl/Pommern (OAW)

Foundation

The company was founded on April 27, 1914 as "Ostdeutsche Albatroswerke GmbH, Schneidemühl". Dr. Walter Huth and engineer Otto Wiener are considered to be the founders. On October 18, 1917, the company became the property of the parent company of "Albatros Gesellschaft für Flugzeugunternehmungen mbH", Johannisthal, by way of purchase and was since then considered a branch of Schneidemühl. The "Militärfliegerschule GmbH, Schneidemühl" was listed as its subsidiary, which was affiliated with the parent plant as a separate "Fliegerschule Department" from March 31, 1918.

Having an aircraft factory, or rather a repair facility, in the east proved to be imperative in order to supply the eastern front with fresh warplanes.

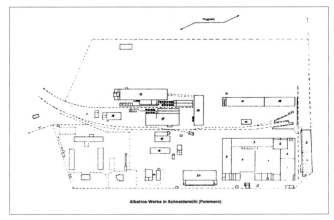

Above: Albatros facilities at Schneidemühl/Pommern.

Factory Facility:

The factory premises were located about 2.5 km from the center of the town of Schneidemühl.

After the company was founded, production was initially started in the attached pilot school (military aviation school)

Above: Map showing the location of the Albatros factory at Schneidemühl/Pommern.

Albatros

Above: Panoramic view of Schneidemühl from 1000 m.

and after completion it was moved to the newly constructed buildings. The factory buildings and workshops were purpose-built.

The factory consisted of a number of timber framed buildings, offices, warehouses, hangars, assembly hangars, aircraft hangars and various production facilities necessary for aircraft construction, which developed as follows:

Year	Production area [m²]
1915	12,300
1916	16,300
1917	21,600

Part of the company was an airfield of about 150 ha near the factory, which consisted only of a sand runway.

Fabrication:

The OAW was engaged in the production, but mainly in the repair of aircraft and their spare parts.
The Albatros-Werke in Schneidemühl was able to achieve a monthly production of 220 aircraft. It produced about 5,000 "Albatros" aircraft of the following types during the war:
a) Category B: Alb B.I, Alb B.II, Alb B.III.
b) Category C: Alb C.I, Alb (Oaw) C.II, Alb C.III, Alb C.V, Alb C.VI, Alb C.X, Alb C.XII
c) Category D: Alb D.III, Alb D.Va, D.VII
d) Category G: Staaken R.VI.

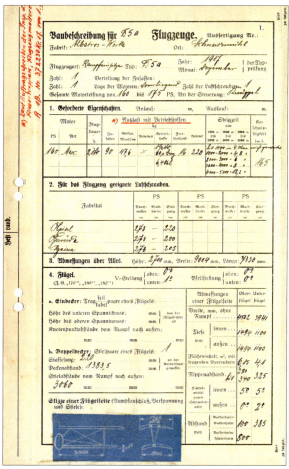

Right: Construction description for Albatros D.V(OAW).

Albatros Aircraft Built at Schneidemühl/Pommern

Type (All Army Aircraft)	Engine [hp], Manufacturer	Crew	Armament	License Built By	Year	Notes
Alb B.I(OAW)	100, Merc. D.I	2	–		1914	
Alb B.II(OAW)	100, Merc. D.I 110, Benz Bz.II 120, Argus As.I	2	–		1914	>250
Alb B.III(OAW)	130, Merc. D.III	2	–		1915	~150
Albatros (C.I OAW) (L7)	150, Benz Bz.III 160, Merc. D.III	2	1 MG		1915	7
Albatros (C.II OAW) (L13)	220, Merc. D.IV	2	1 MG		1915	2 prototypes
Alb C.III(OAW)	150, Benz Bz.III 160, Merc. D.III	2	2 MG		1915	300
Alb C.V(OAW)	220, Merc. D.IV	2	2 MG		1916	25
Alb C.VI(OAW)	180, Argus As.III	2	2 MG		1916	
Alb C.X(OAW)	260, Merc. D.IVa	2	2 MG		1917	200 ordered
Alb C.XII(OAW)	260, Merc. D.IVa	2	2 MG		1917	150
Alb D.III(OAW)	160, Merc. D.III	1	2MG		1916	840
Alb D.Va(OAW)	170, Merc. D.IIIa	1	2MG			600
Fok D.VII(OAW)	180, Merc. D.IIIa	1	2 MG		1918	1300 orders
Alb G.I	4x120, Merc. D.II	3	–		(1915)	1
Staak R.VI(OAW)	4x260, Merc. D.IVa	7–9	4 MG, 1000 kg bombs		1917	3 (4)

Staff and Designers:

Since no new aircraft designs were expected from this subsidiary plant, there were neither design offices nor well-equipped tooling laboratories or test stands.

The workforce grew from the beginning (1915) to about 2,000 by the end of the war.

Since, as already mentioned, no own designs were made, but only built by the parent company or licensed machines, a chief designer cannot be named.

Aircraft Development:

Since the main focus of the OAW was on aircraft repair, it is not possible to speak of actual flight engineering development. Production orders came mainly from the Berlin-Johannisthal headquarters. Other aircraft types were manufactured under license.

The first new machines were school aircraft of the type Alb B.II and B.III were produced. These were then followed by the types Alb C.I, C.III, C.V, C.VI, C.VII, C.X and C.XII. From type C.III on, these aircraft were equipped with fixed machine guns. In 1917/18, fighter single-seaters, type Alb D.III and D.Va were released as new designs. The company produced about 800 of this first type.

In mid-May 1917, the construction of giant aircraft license "Flugzeugwerk Staaken" (somewhat later renamed "Zeppelin-Werke") was started and by March 1918, three R aircraft had been delivered.

The beginning of 1918 also saw the start of licensed production of the Fokker biplane Fok D.VII, and was joined by a light C aircraft of the type Alb C.XV.

Production Supervision (Bauaufsicht): #12

Die Bauaufsicht Nr. 12 in Schneidemühl wurde am 18. September 1916 nach Auflösung der sogenannten „UK" errichtet. Die Bauaufsicht bestand, wie üblich, aus leitenden Offizieren, Unteroffizieren, Soldaten und Zivilangestellten. 3 Nachflieger wurden verpflichtet.

War Material Found by the Inter-Allied Commission of Control (IACC) During Inspections after the End of the War:

At the end of the war, almost the entire inventory, including aircraft and engines, was moved to the main plant in Johannisthal.

Albatros

Situation after WWI:
After the armistice, the main plant "Albatros" in Schneidemühl had new buildings built for the establishment of a foundry. From 1920, 75 workers produced agricultural machinery (mowers and harvesters) and furniture (beds, tables, cabinets).

Above: Alb G.I (1915).

Above: Alb B.I(OAW) (1914).

Above: OAW (Alb) C.I (1915)

Above: OAW (Alb) C.II (1915)

Above: Alb C.III(Oaw) (1915)

Above: Alb D.III (Oaw) (1916).

Above: Alb. D.Va(OAW). (1917)

Above: Fok D.VII(OAW).

Development of Aircraft Factories

Above: Albatros Breguet (1912)

Above: Albatros MZ2 used for W/T trials (1913)

Above: Albatros "Antoninette" (1910)

Above: Albatros Wahl flying boat (1912)

Albatros

Development of Aircraft Factories

4.2.4 Albatros-Militär-Werkstätten, Warschau, Kolejowa (Refla)

Foundation:
The company was a subsidiary of Albatros-Gesellschaft für Flugzeugunternehmungen mbH in Berlin-Johannisthal and was established on February 20, 1916, a year after the German army occupied Warsaw. The Albatros aircraft factory (Refla Militarwerkstätten-Albatros) was established at Mokotow Airport, while the company was called Albatros Militär Reparatur Werkstätten or REFLA Militärwerkstätten, later renamed Reparatur Flugzeugabteilung (Refla) or Reparatur Flugzeug Abteilung (REFLA), i.e. a military repair division.

Factory Facility:
The factory was located in a railroad workshop hall on Kolejowa Street and employed 500 workers. The numerous crew and middle supervision were composed of Poles, including Eng. Zbigniew Arnd (later technical manager of the Central Aviation Workshops).

Aircraft production took place in factory premises rented from the Bormann & Schwede company. The Russian airfield Mokotov served as the airfield, on which the company built 3 hangars. The hangars themselves had a floor area of between 1,600 and 2,400 m2.

Fabrication:
In the beginning, the company dealt with the repair of aircraft and later, as it was not working at full capacity. The plant produced 157 (from an order of 200) Albatros B-II training aircraft and carried out
1916   192 repair airplanes,
1917   267 repair airplanes,
mostly of B- and C-class Albatros aircraft.

Staff and Designers:
The Albatros branch workforce consisted primarily of Polish workers, who were instructed by German foremen from the

Above: Map showing location of Albatros-Militär-Werkstätten, Warschau, Kolejowa (Refla).

Albatros Aircraft Built at Schneidemühl/Pommern

Type (All Army Aircraft)	Engine [hp], Manufacturer	Crew	Armament	License Built By	Year	Notes
Alb B.II/IIa(Refla)	100, Merc. D.I 110, Benz Bz.II 120, Argus As.II	2	–		1917	100

main plant. The total number of employees at the end of 1917 was slightly more than 510.

Aircraft Sevelopment (Since the Outbreak of World War I):

The company did not have its own design office, as mainly repairs were carried out. Later licensed production did not require its own design engineers.

The reason for the establishment of the Albatros military workshops in Warsaw can be traced back to the realization of the Inspectorate of Air Forces (Idflieg) to have a repair workshop in Warsaw quasi close to the Eastern Front to relieve the railroad transports. The plan called for repairing between 40 to 60 aircraft a month.

Preliminary work began in December 1915. Operations began in February 1916, so that the first repair planes could be delivered in the month of April. Since the receipt of repair aircraft did not meet expectations, the company had to be kept busy with the production of new aircraft on the side.

War Material Found by the Inter-Allied Commission of Control (IACC) During Inspections after the End of the War:

No inspection by the IACC took place at this branch plant. Thus, it could not be determined whether military material was still present after the end of the war.

Situation after WWI:

As a result of the armistice on the Eastern Front and the resulting further reduction in the supply of repair aircraft, the company was dissolved before the end of the war after the completion of the last new aircraft. The plant was liquidated in late 1917 and early 1918. The plant's professional staff moved to the workshops of the German officer observer school at Mokotow airport. Employees of the Refla Militärwerkstätten-Albatros factory formed the core of the staff of the Central Aviation Workshops established in Warsaw in November 1918 after Poland regained its independence.

The Polish Alb C.II/IIa were used until end of the 1920s.

Endnotes:

1 Enno Walther Huth (* October 8, 1875 in Altenburg; † May 31, 1964 in Frankfurt am Main) was a German officer, natural scientist and pioneer of the aviation industry.
2 Simon Brunnhuber (* 30. Mai 1884 in Mering; † 6.

Above: Alb B.II(Refla) (1916).

Februar 1936 in Berlin-Johannisthal) was a German Chauffeur, Ingenieur, Fluglehrer und Luftfahrtpionier. 1910 was the first Fluglehrer der Fliegerschule Döberitz. Brunnhuber war Flugschüler von Hubert Latham. Sein PilotenzeugnisNr. 20 erhielt er am 6 August 1910.
3 Eugen Hubert Walter Wiencziers (* March 20, 1880 in Golkowitz, Rybnik District, Silesia; †October 30, 1917 near Speyer) was a German engineer, aviation pioneer, and an Alter Adler. He received his pilot licence No. 8 on May 7, 1910.
4 Wiener previously worked for Rumpler, which initially achieved success in particular through the construction of the Etrich-Tauben.
5 Roger Sommer (August 4, 1877 in Pierrepont, France - April 14, 1965 in Sainte-Maxime) was a French aviator. Born to Belgian industrialist Alfred Sommer, Roger Sommer became involved in aviation at an early age. He broke the record for flight duration in 1909. After that, Sommer started building airplanes. He built 182 airplanes, making him a pioneer in the field. Sommer was a friend of Roland Garros.
6 Benno König (* June 16, 1885 in Untermenzing; † July 1, 1912 in Altona) was a German locksmith, chauffeur and aviation pioneer. He earned his pilot licence No. 45 on December 29, 1910, from Simon Brunnhuber at Johannisthal Airfield, flying a Farman biplane. He then worked as a flight instructor at the Luftverkehrsgesellschaft (LVG), where he trained seven student pilots as early as the spring of 1911. On July 10, 1911, he won the first Deutschlandflug, a cross-country competition over 13 legs.
7 Ernst Heinrich Heinkel (* January 24, 1888 in Grunbach (Schorndorf); † January 30, 1958 in Stuttgart) was a German engineer, aircraft designer, and factory manager. E. Heinkel became known especially after WWI with his own aircraft company.

Above: Staaken R.VI(OAW) "Giant" (1917) on the *Rfa* airfield.

Above: Staaken R.VI(OAW) "Giant" (1917).

4.3 Allgemeine Elektrizitäts-Gesellschaft, Flugzeugfabrik, Hennigsdorf (AEG)

Foundation:

In January 1910, the AEG management decided to start building aircraft. To this end, a "Flight Technology Department" was set up, which was attached to the AEG factories in Hennigsdorf. Locomotives were the main product manufactured at this site.

Initial flying experience was gained with a purchased Wright aircraft, which then led to the construction of a number of experimental types under the direction of Chief Engineer Stumpf.

The workshops required for this purpose were set up in the rooms of an old sawmill located on the AEG site in Hennigsdorf. In the years that followed, the factory premises required for aircraft construction were continuously expanded.

Factory Facility:

Aircraft factory of A. E. G. was adjacent to the large factory for electric locomotives, which this company founded in Hennigsdorf.

The factory had an airfield (see drawing) located a few kilometers from the workshops. In addition, because of the proximity to the Havel River, experiments with seaplanes could be carried out.

Fabrication:

During WWI were manufactured:

Type	Quantity
C	658
J	609
G	523
G (After the armistice and in 1919)	19
Total	1.809

The most important aircraft types have been:
Engine production: A.E.G. Turbinen Fabrik, Hüttenstrasse 14, Berlin manufactured compressors (turbochargers) for aircraft engines.

Staff and Designers:

The number of employees has increased twelvefold since the company was founded, and the area of the site has grown to 70,000 m², 36% of which is covered with buildings.

The responsible technical manager was initially the chief engineer Stumpf, then from 1914 for many years the engineer Georg König.

Aircraft Development:

In 1910, at the instigation of Mr. Baurat Jordan, a "Flight Technology Department" was added to the "A.E.G.". The first flying experiences were gained on a purchased Wright

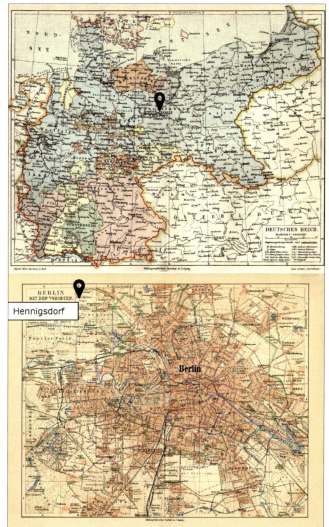

Above: Map of AEG in Hennigsdorf

machine, which then led to the construction of a series of test types under the direction of the chief engineer Stumpf. This resulted in an airplane whose main components, such as the fuselage, undercarriage, wing spars and struts, were made of steel. The special feature of this aircraft was that the wings were retractable on the ground. It was therefore

Right: Advertisement 1913.

Development of Aircraft Factories

Above: Advertisement 1916.

Above: AEG Works wing production.

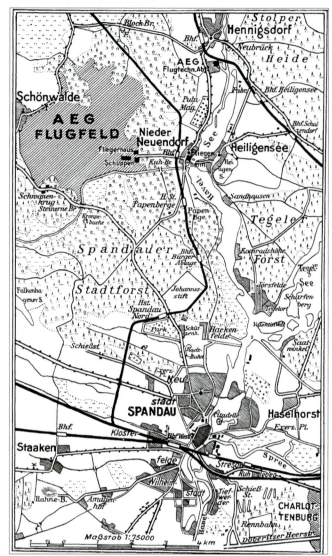

Above: Map of AEG in Hennigsdorf

Above: AEG works (Flugtechnische Abteilung) production hall.

possible, without dismantling, to transport the aircraft on a country road and to store it in small tents in the field.

This resulted in the first military aircraft, the Aeg B.I, in which, in contrast to aircraft from other aircraft factories, all structural parts, such as the fuselage, landing gear, wing spars and stalks, were made entirely of steel, thus guaranteeing weather resistance. The wings were arranged so that they could be pivoted simply by turning them.

Initially, the aircraft were flight tested at the Teltow airfield, as the AEG site in Hennigsdorf was not suitable for this purpose. Since the long distance between the workshops and the airfield led to various inconveniences, AEG decided in the summer of 1912 to set up its own airfield and flying school in the immediate vicinity of Nieder-Neuendorf.

In 1913, the first officers were trained on the new A.E.G. aircraft at the Nieder-Neuendorf airfield; at the same

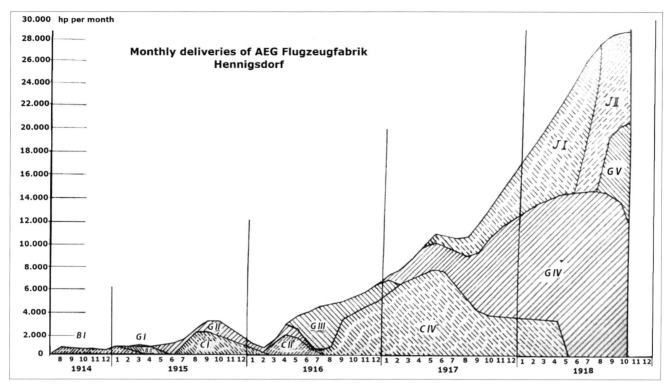

Above: AEG monthly deliveries per aircraft class.

time, Captain Mackenthun[1] joined the flight engineering department in that year.

AEG was the only company that decided to introduce metal aircraft construction right from the beginning. This was based on experience in other branches of industry, where wooden construction was replaced by steel construction in the course of development. Particularly during the war in the construction of large aircraft, these principles proved their worth to a large extent.

By the spring of 1914, the tests had been completed to such an extent that the military administration ordered and accepted a number of the new aircraft. The experience of the Fliegertruppe soon brought the certainty that the steel construction, which had not been used in aircraft construction until then, was particularly well suited to field requirements.

Parallel to the development of land aircraft, construction of seaplanes began in early 1914. The first of this type, the Aeg S.1, was a twin-float biplane powered by a 150 hp Benz Bz.III.

In addition to this unsuccessful attempt, the A.E.G. factory in Hennigsdorf also built a flying boat in 1914.

After the outbreak of war, further series of this type of aircraft were delivered to the air forces. Although here, too, the design did not give rise to any complaints, and the folding of the wings frequently proved advantageous in warfare, increased climbing ability was required for war-related reasons, which is why a better climbing, lighter machine was designed.

This type was built in the autumn of 1914/15 according to the specifications of the engineer Georg König, who from then on was assigned the management of the design office, while maintaining the metal construction and the folding capability of the wings, and was delivered to the field as the Aeg B.II with a 120 hp Mercedes D.II engine, where it proved itself as a replacement for the earlier types due to the 1,000 m increase in climb performance. In terms of aircraft development, it is interesting to note that this type was the first to undercut the weight of the competing wooden aircraft by means of steel construction. This eliminated a prejudice that had hitherto existed against the use and usefulness of steel aircraft. The introduction of aluminum alloys into aircraft construction further displaced the wooden construction method.

The same type was later rebuilt as a C-airplane and introduced into the air forces in larger numbers in the summer of 1915 as the Type Aeg C.I with a 150 hp Benz engine, where it was not inferior to the types of other manufacturers in terms of aeronautical performance.

In the winter of 1914/15, at the suggestion of the Inspectorate of German Air Forces, AEG engineers worked on a large aircraft that was also designed entirely of steel. Initially, the propellers were located at the front, in contrast to aircraft built later by other companies. This method of mounting the engines made for smooth takeoff and comfortable flying. The two 100 hp Mercedes engines located between the engines in the G.I type were connected to the fuselage and undercarriage by a structurally sound tubular

Development of Aircraft Factories

AEG Aircraft Built at Hennigsdorf						
Type (All Army Aircraft)	Engine [hp], Manufacturer	Crew	Armament	License Built By	Year	Notes
Landplanes						
Aeg B.I (Z6)	100-120, Benz, Merc., NAG	2	–		1914	20 ordered
Aeg B.II (Z9)	150, Benz Bz.III 120, Merc. D.II	2	–		1915	10
Aeg B.III (Z10)	160, Merc. D.III	2	–		1915	1
Aeg C.I (KZ9)	150, Benz Bz.III	2	1 MG, small bombs		1915	70
Aeg C.II	150, Benz Bz.III	2	1 MG, 4x10 kg bombs		1915/16	Incl. C.IIa-C.IIe
Aeg C.III	150, Benz Bz.III	2	1 MG		1916	658
Aeg C.IV	160, Merc. D.III	2	2 MG	Fok (incl. C.IVa(Fok)): 300	1916	C.IVvR – lengthened fuselage
Aeg C.V	220, Merc. D.IV	2	2 MG		-, (1916)	
Aeg C.VII	160, Merc. D.III	2	2 MG		-, (1916)	
Aeg C.VIII	160, Merc. D.III	2	2 MG		-, (1917)	
Aeg C.VIIIDr	160, Merc. D.III	2	2 MG		-, (1917)	Triplane, also Aeg PE
Aeg D.I	160, Merc. D.III	1	2 MG		1917	
Aeg Dr.I	160, Merc. D.III	1				1 prototype
Aeg DJ.I (3-D)	195, Benz Bz.IIIb	1	2 MG, 4 small bombs		-, (1918)	1
Aeg DJ.I (2-D)	195, Benz Bz.IIIb 245, Maybach Mb.IVa	1	2 MG, 4 small bombs			2
Aeg J.I	200, Benz Bz.IV	2	1 or 2 MG		1917	~ 400
Aeg J.II	200, Benz Bz.IV	2	1 or 2 MG		1917	> 800
Aeg J.III	260, Merc. D.IVa	2	–		-, (1918)	1
Aeg K.I	2 x 100, Merc. D.I	2				
Aeg G.I	2 x 100, Merc. D.I	3	2 MG		1915	1
Aeg G.II	2 x 150, Benz Bz.III	3	2 MG, 200 kg bombs		1915	15
Aeg G.III	2 x 220, Merc. D.IV	3-4	2 MG, 240 kg bombs		1916/17	45
Aeg G.IV	2 x 260, Merc. D.IVa	3-4	4 MG or 2 MG + 240 kg bombs		1917	542, G.IVa, G.IVb
Aeg G.IVk	2 x 260, Merc. D.IVa	3	2 MG, 2 cannons, bombs		1918	5
Aeg G.V	2 x 260, Merc. D.IVa	3-4	2 MG, 1000 kg bombs		1918	5
Aeg N.I	150, Benz Bz.III	2	1 MG, 300 kg bombs			

AEG Aircraft Built at Hennigsdorf (continued)

Type (All Army Aircraft)	Engine [hp], Manufacturer	Crew	Armament	License Built By	Year	Notes
Landplanes						
Aeg R.I	4 x 260, Merc. D.IVa				1918	2 (7), Test aircraft, crashed in 1918
Seaplanes						
Aeg Flying boat	150 Benz Bz III	2	-		1914	1 prototype
Aeg S1 (Z-5)	150, Benz Bz III	2	-		1914	1 prototype

structure, which reduced engine vibration to an acceptable minimum, since the entire mass of the aircraft resisted vibration. This solution also kept the wing system vibration-free. These design features resulted in high operational reliability. The pilot's seat was located in the center behind the propellers, had a seat for an attendant next to it and a connecting walkway to the balcony, which was intended to prevent accidents during bad landings. This type G.II, which was equipped with two 150 hp Benz engines, was first flown by pilot Kanitz. This giant aircraft was taken into troop service under the designation Aeg G.II in the summer of 1915, where it achieved remarkable success and proved itself as a fighter and bomber. With a wingspan of 16 m, the large aircraft was also the smallest of all competing aircraft and posed fewer difficulties in terms of packaging and accommodation, which was achieved by folding the engines back against the fuselage and thus allowing them to be accommodated in the cargo profile. Road transport was also accomplished without significant disassembly, and stowage in a narrow hangar was made possible by simply folding down the wings.

The constantly increasing demands on the company's ability to deliver could only be met with the support of the sister factories "Maschinenfabrik Brunnenstraße" and the "Turbinenfabrik". Nevertheless, an expansion of the production area was essential. In July 1915, the aircraft department moved into much larger, new premises. At the same time, the aircraft department was transformed into an aircraft factory and placed under the management of Director Baßler, who was assisted by Mr. G. König as chief designer and Mr. Theodor Schauenburg as chief pilot.

Initially, a larger mechanical workshop was attached to the factory in order to become independent of subcontractors by means of its own modern machine tools. Thus, equipment was created that enabled the production of entire fittings from a single piece of material; the profiled tubes were

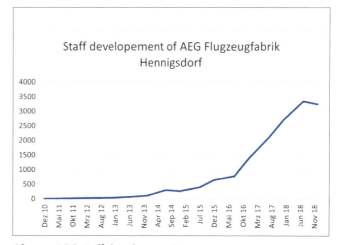

Above: AEG staff development..

Above: First AEG aeroplane with 75 hp Körting 8SL116 (V8) built in 1910.

Above: Second AEG aeroplane from 1911 with 95 hp NAG Type F3 engine.

Above: AEG's Fliegerheim on its own airfield (1913).

Above: Folding wings on a AEG B.I.

Above: AEG flying boat powered by Benz Bz.III (1914).

Above: AEG S.I (Z-5) 1914.

produced from round tube in special drawing equipment, thus simplifying stock-keeping. Welding machines adapted to the individual workpieces made it possible to make the quality of the welding work less dependent on the skill of the welding personnel.

In response to the increased demands for the greatest possible climb rate, a modification of the C aircraft was undertaken. The weight of the AEG C.I was substantially reduced to a permissible level. The new type came to the front as the AEG C.II and certainly satisfied the expectations for climbing.

As an experiment, the C.III type was built, in which the fuselage was designed so that the pilot and observer could see over the upper wings. (Illustration). The performance was essentially the same as that of the C.II aircraft, so the design was not pursued further.

Partly to take advantage of the improved manufacturing possibilities, partly to meet the increased demands of the front, extensive trials were undertaken in the winter of 1915/16, resulting in the Type AEG C.IV, which, equipped with a 160 hp Mercedes engine, was delivered to the fronts in large numbers. Compared with the earlier machines, this type had no welded fittings, but only turned ones, machined from solid; likewise, all considerations for increasing operational safety and durability had been taken into account. For this reason, the ability of the wings to pivot was dispensed with. Not only was it possible to deliver a very large number of this type to the tropics in the summer of 1917, but it was also possible to license the construction of larger series as training aircraft.

Parallel to the C types, the G-aircraft were developed. The AEG G.II received 220 hp Mercedes engines and was delivered to the Fliegerkräfte as Aeg G.III (1916), where they proved themselves especially as bombers. The meanwhile perceived disadvantages of this design led to a completely new design of this type, which, subsequently designated G.IV with two 260 hp engines, served well in the force.

The urgent desire for even better climbing ability brought about the type AEG G.IVb as an additional variant, in which the wingspan, instead of 18 m for G.IV, was now increased to 24 m. Experience at the front, on the other hand, showed that the controllability and maneuverability of the smaller G.IV was preferred to the cumbersome one with greater climbing ability.

The need for bombers led to a modification of the proven AEG C.IV design by enlarging the wings as a night plane. After this modification, it showed particularly good climb performance.

In the spring of 1917, aa AEG D.I fighter was also built on an experimental basis, which, in addition to very good climbing performance, developed a horizontal speed of 210 km/h. As a special feature, it showed wings with a single spar, which reduced the number of bracing cables to a minimum; nevertheless, the strength remained sufficiently high. The same type was designed as a triplane in order to analyze the

Above: Several AEG biplanes (B.I) Z3 and Z6 saving space in an AEG hangar.

Above: Entrance gate of the AEG aircraft factory in Hennigsdorf.

Above: AEG plant in 1918.

Above: AEG G.II with two 150 hp Benz Bz.III engines installed.

climb and speed conditions.

New requests from the front for an armored aircraft were met by converting the AEG C.IV, which was equipped with a 200 hp Benz engine and complete 400 kg armor plating of the fuselage (including engine), was delivered to the field in larger numbers from 1917 under the designation AEG J.I and had now been replaced by a J.II variant. Balanced control surfaces facilitated takeoff and landing, and this change also resulted in better maneuverability.

Meanwhile, another series of aircraft types were created on an experimental basis, intended as light C airplanes, of which type AEG C.VII met the higher climb rates required in January 1917. The same type was also built with a single spar for study purposes to increase speed, and flown as the AEG C.VIII, achieved a speed of 190 km/h. As a triplane, very good climb rates were additionally achieved.

In the meantime, the armored J-planes had proven the high value of armor and caused a desire in the troops to also armor G-planes and equip them with machine guns to be able to engage ground targets, such as tanks. In this way, the AEG G.IVk type was born.

Similarly, an attempt was made to create an armored single-seater to attack the low-flying J planes. The fuselage of this aircraft consisted of an armored hull housing the engine, gasoline tanks, machine guns and pilot. Attached to this was an aluminum tail so that gunfire and flak hits did not damage the power-bearing structure of the aircraft. To increase speed, the same type was also designed as a biplane, in which all bracing was eliminated to achieve greater gunnery safety.

The constant demand of the front for better-rising and more load-bearing bombers was satisfied by the development of the AEG G.V, which, loaded with a 1,000 kg bomb and

Above: AEG G.II with folded wings. Folding wings were required to transport the aircraft on rails.

Above: Rail transport of an AEG G.II with removed wings.

Above: Road transport of an AEG G.II.

Above: AEG G.IVk (k=shortened wings), additionally equipped with 2 X 20 mm Becker cannon.

Above: The AEG D.I reached 210 km/h in level flight.

Above: Railroad transport in front of the AEG shipping hall.

Above: AEG G.IVb with larger wingspan of 24 instead of 18 meters.

Above: Tests with air baffles on an Aeg G.IV did not lead to an increase in performance.

carrying a total payload of 2,100 kg, reached an altitude of 4,000 m with two 260 hp engines, thus fully satisfying the wishes of the front.

Above: As a result of the installation of AEG compressors, the peak altitude of 6000 m was reached during a test flight of a G.IV aircraft with 8 people on board.

AEG

Above: A civilian AEG J.II of Lufthansa postwar for passenger ferries. The typical swan insignia was placed on the fin.

The last investigations on armored aircraft brought the realization that most of the aircraft could not continue the flight due to radiator damage. A.E.G., based on these findings, made the proposal to use engine power to make the air flow through a more inboard radiator. The other positive effect was the reduction of air resistance.

As already mentioned, a complete oxygen production plant and an acetylene plant were added to the aircraft factory in order to make it independent of raw material supplies. Particularly noteworthy is the establishment of its own tube drawing shop, which was created to be able to manufacture the spars and axles, which are made of high-quality steels, in-house. Tests showed that it was possible to manufacture tubes from nickel-free material with the same strength properties as the chrome-nickel tubes. Similarly, it has been possible to produce tubes of rectangular cross-section from the same material, the narrow side of which has a thicker wall than the long side, in order to make the best use of the material for wing spars.

A major disadvantage of piston engines in general is the loss of power with increasing altitude due to reduced air density. A compressor developed by A.E.G. was intended to remedy this situation by supplying compressed air to the engine. The performance of the engines, especially at altitude, increased enormously.

War Material Found by the Inter-Allied Commission of Control (IACC) During Inspections after the End of the War:

42 different aircraft
44 aircraft J.II and G.V
114 aircraft engines
8 compressors

Situation after WWI:

After the armistice, AEG converted its Type G.V into a transport aircraft for 10 passengers. In 1919, AEG had then discontinued its aircraft factory as a result of the Versailles Treaty and the associated manufacturing restrictions. The factory was incorporated into the production of the parent factory in Hennigsdorf (electric locomotives, agricultural machinery, etc.).

Likewise, the AEG factory in Hüttenstrasse had not been active in aviation since 1920. Here, as a result of the special knowledge available, it had specialized in the manufacture of high-performance steam turbines.

Above: AEG B.I (Z6) (1914)

Above: AEG B.II (Z9) (1915)

Above: AEG B.II (Z9) with Daimler (1915).

Above: AEG B.III with 120 PS Daimler (1915)

Development of Aircraft Factories

Above: AEG C.I (KZ9) (1915)

Above: AEG C.II (1915/16)

Above: AEG C.III (1916)

Above: AEG C.IV (1916)

Above: AEG C.V (1916)

Above: AEG C.VII-3 (1916)

Above: AEG C.VIII (1917)

Above: AEG C.VIIIDr (1917)

AEG

Above: AEG D.I (1917)

Above: AEG Dr.I (1917)

Above: AEG J.Ia (1917)

Above: AEG J.II (1917)

Above: AEG DJ.I triplane (1918)

Above: AEG DJ.I biplane (1918)

Above: AEG K.I (1915)

Above: AEG G.I (1915)

Development of Aircraft Factories

Above: AEG G.II, triple tail (1915/16)

Above: AEG G.III (1916/17)

Above: AEG G.IV (1917/18)

Above: AEG G.IVb (1918)

Above: AEG G.IVb-Lang (1918)

Above: AEG G.IVk (1918)

Above: AEG G.IVk (1918)

Above: AEG G.V (1918)

AEG

Above: AEG G.V Civil (1918)

Above: AEG N.I (1916)

Above: AEG R.I (1916/17)

Above: AEG Sea biplane (1914)

Above: AEG Flying boat (1914)

Endnote

1 Walter Mackenthun (* August 17, 1882 in Berlin; † June 7, 1948 ibid) was a German aviation pioneer and co-initiator of military aviation. On March 7, 1911, he earned his pilot's license (No. 72) at the Döberitz Flying School and became a lieutenant. In World War I, from 1914 to 1917, he was a captain and squadron leader of the Habsheim Fighter Command (renamed Jasta 15 in Sept. 1916). From 1917, he and Egon von Rieben were managing directors of the Deutsche Luft-Reederei, founded on Dec. 13, 1917, as the Studiengesellschaft. In mid-August 1918, Mackenthun and von Rieben tried in vain to persuade Hugo Junkers to enter into a manufacturing alliance with the parent company AEG for the planned Junkers F 13.

Above: AEG-Wagner (1914) wing testbed held in the Polish Aviation Museum in Krakow.

Below: AEG-Wagner (1914) wing testbed placard held in the Polish Aviation Museum in Krakow.

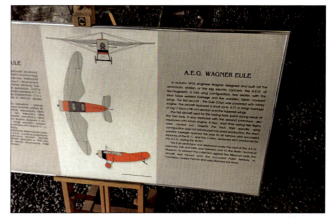

Development of Aircraft Factories

4.4 Automobil und Aviatik A.-G.

4.4.1 Automobil und Aviatik A.-G., Leipzig-Heiterblick (Av)

Foundation:

The company emerged from the bicycle and automobile factory founded by Georg Chatel in Mulhouse i. Alsace in 1900, which was then moved to Mulhouse-Burzweiler[1] in 1906 into a newly built factory building. In 1910, Georg Chatel founded Aviatik-GmbH with another partner, the director Jeannin of Argus-Motoren-Werke, and appointed Julius Spengler, already active in the first company, as its managing director. These two companies were merged in 1911 by founding the "Automobil- und Aviatik-A.-G.".

While around 1900 the original company was engaged only in the construction of bicycles and automobiles, Aviatik-GmbH turned to the construction of aircraft. After merging the two companies, bicycle construction was separated from the former and only automobile production, which dealt in particular with the construction of car bodies, was retained, while the operation of the second company, the construction of airplanes, was continued.

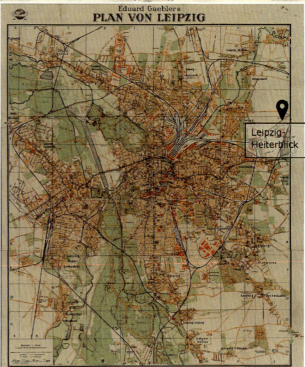

Above: Map of the German Reich and Liepzig.

Above: Advertisement in *Motor*. (1913)

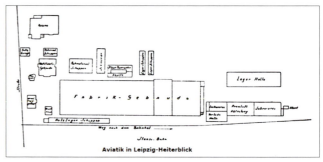

Above: AEG-Wagner (1914) wing testbed held in the Polish Aviation Museum in Krakow.

Aviatik

Factory Facility:
At the outbreak of World War 1, as Burzweiler was too close to the enemy border, the company was moved to Freiburg im Breisgau until it was finally able to relocate to the newly built and modernly equipped Leipzig-Heiterblick factory site in June 1916. The flying school operated at the Habsheim airfield near Mulhouse was also moved to Freiburg at the beginning of the war, and then in 1915 it was relocated to Mockau near Leipzig on the grounds of the Leipzig Luftschiffhafen-A.-G. At the end of 1916, the flying school burned down completely. This fire, in turn, gave rise to the purchase of the insolvent Grade-Werke in Bork (Mark), where the Aviatik company subsequently established a branch, which was primarily concerned with school operations, but then also with the construction of school machines and repairs.

The factory in Leipzig-Heiterblick, which was opened in the fall of 1915, had a volume of about 325,000 m³ and a built-up area of about 25,000 m². With its extensions, the plant was completed at the end of 1917, although most of its operations had already started in 1916. The facility itself was designed to accommodate 1,000 to 1,200 medium aircraft. It was located 10 minutes from Heiterblick Station, had its own railroad siding, and was situated on a newly constructed road.

In essence, the factory complex consisted of an administrative building, which housed the director's offices and commercial offices, a factory building (except for a few square meters, only one-story) with its associated boiler house for heating the factory premises, because an overhead line supplied the necessary electricity.

Fabrication:
The Automobil und Aviatik A.-G. in Leipzig-Heiterblick built[2]:

Type	Number
B	246
C	2.259
D	80
G	10
R	9
Total	2.604

Staff and Designers:
The number of workers, which had been about 200 until the factory was moved to Freiburg i.Br. in mid-1914, rose to about 700 there and was increased to about 1,400 after the factory was moved to Leipzig. In order to meet the increased requirements of the army administration, the workforce during the war even reached up to 1,600 workers in the aircraft construction plant.

The company was managed by the directors:
  Vogtenberger: Commercial Director,
  Dipl.-Ing. Hoffmann: Technical Director,
  Victor Stöffler: Operations Manager and Chief Pilot
  Linke: management of the branch office in Bork

Above: Advertisement in *Motor*. (1913)

Designers since foundation were Dipl. Ing. Robert Wild, Dipl. Ing. Cl. Descamps, Ing. Paul Koechlin (seaplane construction), Dipl. Ing. Werner as well as Chief Engineer Rau in R-aircraft construction.

Aircraft Development:
Automobil und Aviatik A.-G. is one of the oldest German aircraft factories. After successfully reproducing French aircraft models in 1910, the company participated in military maneuvers with purely German aircraft developments in the years before the war. This formed the basis for further aircraft orders from the German army administration.

To foreign countries, the company delivered aircraft to Russia and Belgium several times. In other countries, e.g. Italy, Aviatik licenses were used to advance aircraft construction.

The successes achieved by Automobil- u. Aviatik A.G. in October 1913, when Viktor Stöffler won the Grand Prix of the "Deutsche Nationalflugspende" (2,100 km in 24 hours) and in the following year, February 1914, when Karl Ingold won the first prize in the German City Flight (1,600 km in the air),

Development of Aircraft Factories

Automobil and Aviatik A.-G. Aircraft Built in Leipzig-Heiterblick[2]

Type (All Army Aircraft)	Engine [hp], Manufacturer	Crew	Armament	License Built By	Year	Notes
Av B-type (P13, P14)	85-100, Argus, Benz, Mercedes	2	–		1915	113
Av B.I (P15B)	100, Merc. D.I	2	-„-		1914	246
Av B.II (P15A)	120, Merc. D.II	2	-„-		1915	
Av C.I (P25)	150, Benz Bz.III / 160, Merc. D.III	2	1-2 MG	Han: 146	1915	400
Av C.Ia	160, Merc. D.III / 150, Benz Bz.III	2	-„-		1915	1, prototype for Av C.III
Av C.II	200, Benz Bz.IV	2	1-2 MG		1915	75+1
Av C.III	160, Merc. D.III	2	2 MG		1916	250
Av C.IV [DFW C.V]	200, Benz Bz.IV	2	2 MG		1916	prototype
Av C.V	180, Argus As.III	2	2 MG		1916	prototype
Halb C.V(Av)	200, Benz Bz.IV	2	2		1916	150
Av C.VI [DFW C.V]	180, Argus As.III	2	2 MG		1917	1,400
Av C.VIII	160, Merc. D.III	2	2 MG		1918	1
Av C.IX	200, Benz Bz.IV	2			1918	3 prototypes
Av D.I [Halb D.II]	120, Merc. D.II	1	1 MG		1916	~80
Av D.II	160, Merc. D.III	1	2 MG		1916	prototypes
Av D.III	195, Benz Bz.IIIbo	1	2 MG		1917	2 prototypes
Av D.IV & D.V	195, Benz Bz.IIIbv	1	-„-		1917/18	prototypes
Av D.VI	195, Benz Bz.IIIbm	1	-„-		1918	
Av D.VII	195, Benz Bz.IIIbm	1	-„-		1918	
Go G(L).VII(Av)	2x260, Merc. D.IVa	2	-„-		1918	30 (100)
Staak R.VI (Av R.I)	4x260, Merc. D.IVa or 4x245, Mayb. Mb.IVa	7-10	4-7 MG		1918	6 (R33/16 – R35/16, R52/16 – R54/16)
Staak R.XVI (Av R.II)	2x220, Benz Bz.IV / 2x530, Benz Bz.VI	7	5 MG		1918/19	1 (3) (R49/16 – R51/16)

Above: Pilot Stöffler and his Aviatik P15 after the record flight to Warsaw (1913).

Above: Aviatik monoplane (E12) developed in Mulhouse i.E.

Aviatik

Above: Early Aviatik seaplane during a competition at Lake Constance. Center float and floats under the wings.

drew attention to this company and its aircraft designs.

The first "Aviatik" aircraft was a biplane with a rear-mounted engine and propeller. The appearance of this type was similar to the well-known French designs.

For the seaplane competition in Heiligendamm in October 1912, the company redesigned the above-mentioned biplane into a seaplane. In addition to this WDD (wingspan 20 m, 100 hp Argus engine), construction was also completed on the monoplane then based on the Etrich pigeon, on which Jeannin won 1st prize at the Schwabenflug in September 1911.

During the war, Automobil- u. Aviatik A.G. achieved successes that testified to a skilful ability to adapt to the increased and constantly growing demands placed on military aircraft by the army administration in terms of flight characteristics and combat capability.

In 1914/15, the first C airplanes were built. The Av C.I with a 160 hp Mercedes engine and frontal radiator reached a top speed of 142 kph. A striking feature was that the observer sat in front of the pilot. This resulted in a better field

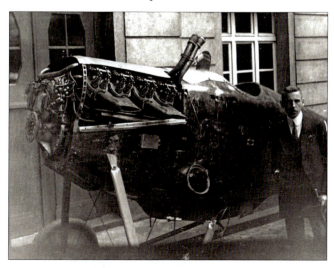

Above: Fuselage of the Av D.IV prototype with built-in V8 engine Benz Bz.IIIbv (195 hp).

Above: Main supplier for Aviatik aircraft in Austria: A. Weisser.

Above: Aviatik C.IX prototype.

of fire. The type Av C.III (built in 1916), also with Mercedes D.III engine, was aerodynamically improved to such an extent that its top speed already reached 160 km/h. This was achieved by reducing the damaging frontal drag, installing a wing radiator and modifying the wing profile and torpedo-like fuselage shape.

The type Av C.V was created in 1917, powered by a 180 hp Argus engine, had a cranked upper wing with

Above: Staak R.XVI (R50/17) built by Aviatik in flight.

flights thanks to a reinforced design.

About 30 large Go.VII(Av) aircraft were built under license from Gotha in 1918. The R-planes (licensor: Zeppelin-Staaken) still completed in 1918 were not used in the war.

Production supervision (Bauaufsicht): # 7
To assist in the procurement of materials and in particular for the acceptance of wartime aircraft, the company was assigned the production supervision of the Inspektion der Fliegertruppen (IdFlieg) No. 7.

War material found by the Inter-Allied Commission of Control (IACC) during inspections after the end of the war (1920):
In the hangars and at the airfield were found 97 aircraft, including a Staak R.XIV(Av) R-aircraft and a civilian aircraft developed from a Staaken R-aircraft and completed after the end of the war;
44 aircraft engines.

Situation after WWI:
In 1920, the aviation factory in Leipzig was financially supported by the Benz & Cie. company. Production was switched to Benz agricultural tractors.

Above: Advertisement 1917.

direct connection to the upper fuselage spar. or to the corresponding transverse frames. The purpose of connecting the upper wing to the fuselage was to create an improved upward field of view and firing.

Im Jahre 1917 entstand das einstielige Flugzeug Av C.VIII mit Stirnkühler. Zur Schaffung eines größeren Flügelabstandes erhielt der Rumpf einen kielartigen Ansatz zur Befestigung der Unterflügel.

In 1917, the Av C.VIII single-aisle airplane with a frontal radiator was created. To achieve greater wing spacing, the fuselage was given a keel-like attachment for mounting the lower wings.

In the class of D airplanes, the Av D.II should be mentioned: 1916 model year, 160 hp Mercedes D.III, hydrofoil radiator. The front part of the fuselage, especially canopy, engine mount and undercarriage were a complete steel tube construction. The driver's seat was mounted on the extended continuous U-beams of the engine foundation.

The type Av D.VII, equipped with the V8 high-speed Benz Bz.IIIbo engine, was capable of looping and swooping

Above: Aviatik P13 (B type) (1913)

Aviatik

Above: Aviatik P14 (B type) (1913)

Above: Aviatik B.I (P15) (1913/14)

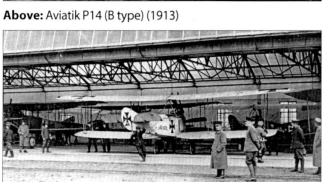

Above: Aviatik B.I I 1329/15)

Above: Aviatik C.I (1915)

Above: Aviatik C.II (1916)

Above: Aviatik C.III (1916)

Above: Aviatik C.IV (1916)

Above: Aviatik C.V I (1916)

Development of Aircraft Factories

Above: Aviatik C.VI (1917)

Above: Aviatik C.VIII (1917)

Above: Aviatik C.IX (1917)

Above: Aviatik D.I (license-built Halberstadt D.II) (1916)

Above: Aviatik D.II (1916)

Above: Geest Fighter (Aviatik D.II modified with Geest wing) (1916)

Above: Aviatik D.III (1917)

Above: Aviatik D.IV (1917/18)

Aviatik

Above: Aviatik D.VI (1918)

Above: Aviatik D.VII (1918)

Above: Go G.VII (Av) (1918)

Above: Halb C.V(Av) (1918)

Above: Halberstadt D.II(Av) (1918)

Above: DFW C.V(Av) (1917)

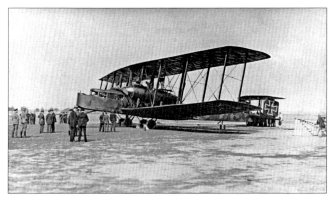

Above: Staaken R.VI(Av) 33/1 (1916)

Above: Staaken R.VI(Av) 52/17 (1917)

4.4.2 Automobil- und Aviatik AG, Subsidiary Bork i. Mark

Above: Label Vignette.

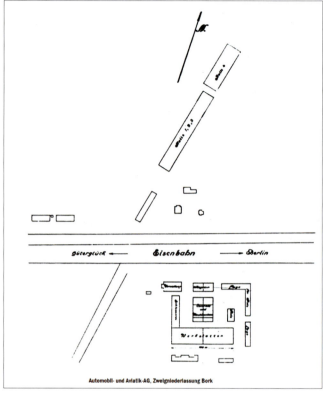

Above: Map Grade plant in Magdeburg.

The Bork branch of Automobil- und Aviatik AG has a differentiated history, having evolved from the "Hans Grade Flieger-Werke".

Hans-Grade-Werke, Bork i. Mark

After Hans Grade[3] completed an engineering course at the Grevenbroich machine factory near Cologne at the age of 20, he began his studies at the Technische Hochschule in Berlin-Charlottenburg, the same university where Otto Lilienthal had previously studied. Equipped with the internship experience, Grade developed engine construction while still a student. The work progressed so well that he soon received three patents for his engine development. An engine-powered bicycle followed. After graduation, Grade returned to Köslin and accepted a managerial position in a small Köslin engine workshop, which he took over as manager with the support of the sponsor Hentschel in 1904. To increase sales opportunities, Hentschel and Grade decided to move to Magdeburg.

On September 23, 1905, Hans Grade signed a shareholders' agreement for Grade-Motorwerke GmbH, was appointed to the board of directors and became managing director. This GmbH manufactured in particular motorcycle engines for the well-known company Burckhardtia[4].

In 1907, Grade had just completed his compulsory military service as a one-year volunteer and began designing the first aircraft engine. At the same time, he manufactures the first parts for an airplane in a workshop of his engine factory. The award of the Lanzpreis der Lüfte (Lance Prize of the Skies) spurred Grade on to accelerate work on his aircraft; after all, the winner received 40,000 marks.

Factory Facility:

In 1909, Georg Rothgießer, a Jewish businessman and enthusiastic aviation engineer, succeeded in persuading Grade to leave the too small airfield in Magdeburg and move to the "Mars" airfield near Bork near Berlin, where other aircraft manufacturers were already in the process of setting up shop.

At the beginning of the new year, several permanent hangars were built, which were approved by the trade police on March 15, 1910. On the wall of one of the hangars is written: "Hans Grade Fliegerwerke". In 1912, about 50 people were employed here, on a total built-up area of 3,000 m². On May 1, 1910, the first aircraft was delivered to the Ikarus airline.

The local conditions changed only insignificantly due to the closure of the Grade works. Some new workshops were added by Aviatik, which were equipped with machines from the parent company. The built-up area increased to 4,200 m² (end of 1917). In addition, 6,000 m² of the pilot school were added.

Staff and Designers:

In total 550 employees were busy with repairs on different types of aircraft and the manufacture of Albatros B.II.

Aircraft Development:

Hans Grade developed a new monoplane from the "Libelle", the "Schwalbe" in the basic models A, B and C, which differed in their wingspans of 8, 8.5 and 12 meters. The C basic type was designed

Aviatik

as a two-seater. Grade also offered a total of seven different engines. This large number of possible combinations was intended to help meet special customer requirements. The competition was huge at the time. Grade felt this during many competitions and display flights in which he participated. The speeds that could be achieved with the "Schwalbe" (swallow) were no longer sufficient, so Grade decided to develop a racing monoplane. Two new models were developed. Type D was a single-seater with a 30 or 45 hp four-cylinder engine. For the first time, this type had a partially aluminum-clad fuselage made of tubular steel, in which the pilot was accommodated. However, the attainable maximum speed increased only to 110 - 115 Km/h.

Although further types were also built at the Grade works in 1913 and 1914, the expected success failed to materialize, not least because of the collapse of civilian sport aviation. In particular, the measures taken under the National Flight Donation led to a steady decline in sales.

After the outbreak of the World War, the capacities of Grade Flieger-Werke were used to repair front-line aircraft. Therefor an independent aeronautical development in Bork did not existed after the beginning of WWI. Grade's influence within his company was so weakened by the directive rights of the military construction supervisors that he decided to sell his factory in 1916. The date of the establishment of the above company is January 1, 1917, the date of the takeover of Hans-Grade Flieger-Werke by Aviatik. The company manufactured about 550 Albatros B.II aircraft under license and repaired 250 Aviatik C I, C II and C III and DFW C V aircraft in 1917.

Aircraft Built by Grade in Magdeburg						
Type (All Army Aircraft)	Engine [hp], Manufacturer	Crew	Armament	License Built By	Year	Notes
Alb B.II(Av)	120, Merc. D.II	2	-		1917/18	~220

Endnotes

1 Today: Mulhouse, City district Bourtzwiller (France)
2 According to the report of the Interallied Control Commission (IACC).
3 Johannes Gustav Paul "Hans" Grade (* May 17, 1879 in Köslin, Pomerania Province, Prussia, now Poland; † October 22, 1946 in Borkheide) was a German mechanical engineer, entrepreneur and aviation pioneer.
4 Fritz Burckhard, Magdeburg, Manufacturer of bicycles.

Right & Below: The Aviatik D.VII was the final Aviatik design to be tested, and apparently it was placed in production because the Inter-Allied Control Commission found 50 completed examples after the war. Power was by the 195 hp Benz Bz.IIIbm V-8, which was too late for the war.

Development of Aircraft Factories

4.5 Bayerische Flugzeug-Werke AG, München (Bay)

Foundation:
Bayerische Flugzeug-Werke AG was founded on February 20, 1916. The company (BFW AG) emerged with Bavarian state support from the uneconomically operating Gustav Otto Flugmaschinenwerke in Munich, whose headquarters were located at Oberwiesenfeld. The Otto-Werke facilities at the Munich headquarters were taken over, and the owner Gustav Otto (1883-1926) was paid off.

Factory facility:
The company had its own factory premises of about 35,000 m², with the possibility of establishing an airfield to the north of the factory.

Until the end of the war, however, the test and evaluation flights took place at the Oberwiesenfeld parade ground (Munich-Milbertshofen).

Fabrication:
The Bayrische Flugzeug-Werke AG produced about 1,000 aircraft only as license builds. In addition, repairs were carried out in another former "Deutschland" plant.

In 1916, 389 aircraft were produced.
In the following year, 1917, 1,039 new aircraft were built. In addition, 386 repairs were carried out.

Type	Licence holder
B.II	Albatros
C.Ia/b	Albatros
C.III	Albatros
C.V	Halberstadt
C.VII	Albatros
C.X	Albatros
C.XII	Albatros
CL.II	Halberstadt

Staff and Designers:
The number of employees has grown steadily since the company was founded and since the takeover. While there were 350 employees on February 20, 1916, the number had already risen to around 2,400 by March 1918. In addition to the director, Dipl.-Ing. Peter Eberwein, the chief engineers were Dr. Rippel, Dipl.-Ing. Gaule and Chief Engineer Scheuermann.

The chief pilot was initially Mr. Weyl, and since spring 1918 Mr. Neumaier.

Aircraft Development:
The Bavarian Aircraft Works took over most of the workers and civil servants who were then employed at Otto-Werke Oberwiesenfeld and Werk "Deutschland". By January 1916, about one third of the workers were still employed by Otto-Werke in order to complete the orders still in progress for Otto.

At the insistence of the Central Acceptance Commission (ZAK), both plants were again substantially enlarged in 1917, so that by December, for example, some 1,650 men were employed.

Above: Map of the German Reich and München.

It was now possible to produce 200 new aircraft a month and to complete about 30 aircraft repairs.

The higher number of workers in relation to production can be explained by the fact that, with very few exceptions, Bayerische Flugzeugwerke manufactured all spare parts itself. From the company's founding until December 31, 1917, 1,400 new aircraft were built, and 346 repairs and a larger number of aircraft spare parts were also completed.

Nevertheless, the company also developed and tested its own aircraft types. Initially, this was a light fighter biplane of normal design with an I-strut. A striking feature was the ring front radiator for the Daimler D.III engine. In 1918 the CL.II and CL.III types were tested. A triplane designed as a night bomber, also from 1918, was also only a prototype. The latter had an empty weight of 1,500 kg, the payload was 1000 kg.

A monoplane developed as a shoulder-decker in 1918 also remained an experiment.

BFW

Aircraft Built by Bayerische Flugzeug-Werke AG, München (Bay)						
Type (All Army Aircraft)	Engine [hp], Manufacturer	Crew	Armament	License Built By	Year	Notes
Bay CL.I	160, Merc. D.III	2			1916/17	1 prototype
Bay CL.II	185, Mana III	2			1918	1 prototype
Bay CL.III	200, Benz Bz.IV	2			1918	1 prototype
Bay N.I	260, Merc. D.IVa	2	MG, bombs		1918	1 prototype
Monoplane experimental	160, Merc. D.III				1918	1 prototype
Alb B.II(Bay)= Bay B.I	100-120, diverse	2	-		1916/17	~180
Alb C.Ia(Bay)	180, Argus As.III	2	1 MG		1917	370
Alb C.III(Bay) = Bay C.I	160, Merc. D.III	2	2 MG		1917/18	~500 as Bay C.I, ~150 as trainer
Halb C.V(Bay)	200, Benz Bz.IV	2	1 MG		1918	400 orders
Alb C.VII(Bay)	200, Benz Bz.IV	2	2 MG		1917/18	275
Alb C.X(Bay)	260, Merc. D.IVa	2	1 MG, bombs		1917/18	50
Alb C.XII(Bay)	260, Merc. D.IVa	2	1 MG, bombs		1917/18	180
Halb CL.II(Bay)	160, Merc. D.III	2	1 MG		1918	100
Halb CL.IIa(Bay)	190, Argus As.IIIa	2	1 MG		1918	100

Production Supervisors (Bauaufsicht): #20
This construction supervision consisted of two officers, three foremen, sergeants, and staff, as well as a test pilots.

War Material found by the Inter-Allied Commission of Control (IACC) During Inspections after the End of the War:
92 aircraft
222 engines of different manufacturers.
Miscellaneous spare parts

Situation after WWI:
After the armistice, Bayerische Flugzeug-Werke A. G. rebuilt its machinery and equipment and began manufacturing furniture, agricultural machinery and later gasoline engines. Towards the end of 1920, it employed 100 workers.

At this time, BFW AG also housed the air transport company "Bayerischer Luft Lloyd" in its buildings.

Above: Bay CL.II (1918)

Above: Bay CL.I (1916/17)

Below Left: Bay CL.III (1918)

Below: Bay N.I (1918)

Development of Aircraft Factories

Above: Bay experimental monoplane. (1918)

Above: Albatros B.II(Bay) (1916/17); note wooden wheels.

Above: Albatros C.Ia(Bay) (1917)

Above: Albatros C.III(Bay) (1917)

Above: Albatros C.VII(Bay) (1917/18)

Above: Albatros C.X(Bay) (1917/18)

Above: Halberstadt CL.II(Bay) (1918)

Above: Albatros C.XII(Bay) (1917/18)

Above: Halberstadt C.V(Bay) (1918)

4.6 Bayerische Rumpler-Werke AG, Augsburg (Bayru)

Foundation:
On October 24, 1916, the continuing high demand due to the war led to the founding of a Rumpler subsidiary in Augsburg. One of the investors in Bayerische Rumpler-Werke AG was the entrepreneur August Riedinger with his balloon factory. The "Bayerische Rumpler Werke GmbH" was a kind of branch of the "Rumpler Werke GmbH" in Johannisthal, which, however, always had its own budget and retained its independent management and thus always a certain freedom vis-à-vis the parent company. During the war, it was given the short name "Bayru" by the Flugzeugmeisterei.

Factory Facility and Airfield:
Between the groundbreaking ceremony for the construction of the production halls with adjacent airfield on November 25, 1916, and the completion of the first aircraft on July 1, 1917, only a little more than eight months passed. In May 1917, the company already owned 3 factory buildings totaling 4,900 m². In January 1918, another shipping hall and a warehouse of 2,900 m² each were added. This was followed by an administration building and other buildings, such as a canteen and a civil servants' casino.

At the end of the war, about 350 aircraft were produced in Augsburg under the leadership of plant manager Otto Meyer. In the Berlin plants, the number of aircraft built between 1908 and 1918 was 3060. At the peak of production in 1918, a total of about 3300 workers were employed in Berlin and Augsburg.

The plant was very conveniently located directly on a local rail line and had its own siding. The airfield, with a size of about 100 hectares, directly bordered the workshops.

Fabrication:
The Bayerische Rumpler-Werke AG in Augsburg primarily produced new front-line aircraft under license. Naturally, these included types from the parent company, such as Rumpler's Ru C.IV and Ru C.I. Repairs were carried out to the same extent, so that about 20 new aircraft and up to 30 repair aircraft were delivered each month.

There were 318[1] (430[2]) licensed aircraft produced at Augsburg during the 13 months of its existence.

Staff and Designers:
Since there were no skilled workers available for aircraft construction, all personnel first had to be trained and the workshops set up. In 1918, 2 years after its foundation, just under 800 men were employed.
Since only licensed constructions and repairs were carried out, the Bayerische Rumpler-Werke did not have a separate design office, and thus no design engineers.

Aircraft Development:
No independent aeronautical development took place in this subsidiary of Rumpler-Werke.

The first "Made in Augsburg" aircraft took off on July 1, 1917, on the grass runway adjacent to the factory hangars.

Above: Map of the German Reich and Augsburg.

The "C IV" was a two-seater biplane with a wingspan of 12.66 meters and a 260 hp Mercedes engine. Its top speed was 170 kilometers per hour, and its attainable ceiling was 6000 to 7000 meters. Its tanks held 240 liters of gasoline. This was enough for up to four hours of flight time. The aircraft was a purely military production. As a fighter, the biplane of wooden construction, was equipped with two machine guns. As a reconnaissance aircraft, it was fitted with

Development of Aircraft Factories

Above: View of the Bayerische Flugzeug-Werke in Augsburg.

a series-image camera under the observer's seat. Instead of the camera, the aircraft could carry small droppable explosive devices weighing up to 100 kg.

Production supervisors (Bauaufsicht): # 37
Construction Supervision No. 37 included two officers, three foremen, four enlisted men, and one inspector pilot.

War Material Found by the Inter-Allied Commission of Control (IACC) During Inspections after the End of the War:
21 aircraft
3 Engines

Situation after WWI:
In early 1920, Rumpler-Werke in Augsburg began building agricultural machinery. In May 1920, the company employed about 280 workers in this field.

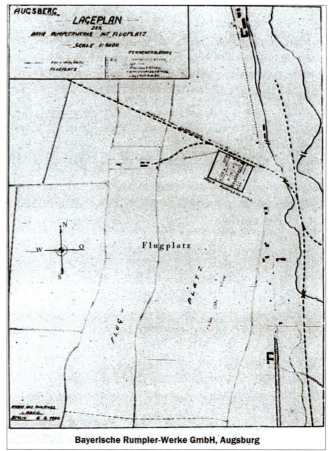

Above: Site plan Bayerische Flugzeug-Werke in Augsburg.

Aircraft Built by Bayerische Rumpler-Werke AG, Augsburg (Bayru)						
Type (All Army Aircraft)	Engine [hp], Manufacturer	Crew	Armament	License Built By	Year	Notes
Ru C.I(Bayru)	160, Merc. D.III	2	1 MG	-	1917	100 orders
Ru C.IV(Bayru)	240, Merc. D.IVa; 245, Mayb. Mb.IVa; 245, BuS BuS.IVa	2	1-2 MG, Bombs or camera	-	1918	In total 350 orders; Aircraft with BuS engines also known as Ru C.IV(BuS)

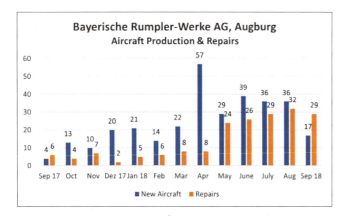

Above: Bayerische Rumpler-Werke AG Augsburg, aircraft production and repairs.

1 See „Rumpler-Zehn Jahre Deutsche Flugtechnik", Ecksteins Biografischer Verlag, Berlin, 1919
2 According to IACC Report

Above: Ru C.I (Bayru)

4.7 Daimler-Motoren-Gesellschaft AG, Stuttgart-Sindelfingen, Abteilung Flugzeugbau (Daim)

Foundation:

The company was founded in Cannstadt in 1890 by Gottlieb Daimler[1]. Aircraft construction in the factory halls in Stuttgart-Sindelfingen, however, did not begin until the summer of 1915 at the behest of the Inspectorate of Aviation Troops. Up to this point, DMG's wartime production had focused primarily on the construction of trucks and engines, including aircraft engines.

The aircraft construction department used the halls of its body shops in Sindelfingen. The Flugzeugmeisterei assigned the abbreviated designation "Daim" to this Daimler aircraft construction department.

Factory Facility and Airfield:

The Daimler factory was built in 1915 on the border of a railroad line outside Sindelfingen. Because of its location, it could be expanded indefinitely. Some of the buildings were made of reinforced concrete, others of brick, most of them of wood.

The company was located at the airfield of the Flieger-Ersatzabteilung (Fea) 10 in Sindelfingen and subsequently owned its own airfield area. The buildings needed to manufacture the aircraft were completed in the spring of 1917.

Fabrikation:

The company was originally engaged in the construction and development of aircraft engines and, since the establishment of the aircraft construction department, in the production of four-engine large-scale warplanes, G-airplanes, the series construction of licensed machines (type G III) and experimental machines of type D, fighter single-seater biplanes with 185 hp V-engines from its own production.

Based on the joint preliminary work the Union Flugzeugwerke (see there), a new large fighter R1 was designed in Sindelfingen in the fall of 1915 under still quite primitive and makeshift conditions under the direction of engineers Baurat Rittberger and Karl Schopper (both formerly employed by Union), and 2 were built by hand with relatively primitive equipment. Like the improved R.II, most of the parts were manufactured by Schiedmayer and then shipped to Sindelfingen for assembly. The R.I's fuselage was more robust than its predecessors, and the unreliable inverted cylinder engines were replaced by the reliable 160 hp Mercedes D.III engines. The R.I's maiden flight took place at Sindelfingen in late 1915.

Aircraft production in March-May 1917 was 3 units, in June-September 5 units each month, and increased to 12 units of Fdh G.III by the end of 1917. By the end of the war, the company reached a monthly production of new Fdh G.III aircraft in the amount of 30 units per month.

Staff and Designers:

The workforce of the aircraft construction department grew

Above: Map of the German Reich and Sindelfingen.

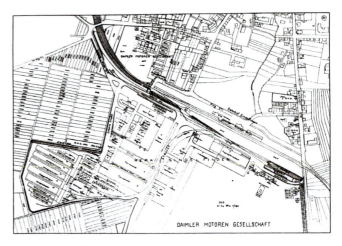

Above: Site plan DMG in Sindelfingen

Aircraft Built by Bayerische Rumpler-Werke AG, Augsburg (Bayru)

Type (All Army Aircraft)	Engine [hp], Manufacturer	Crew	Armament	License Built By	Year	Notes
Union G.I	4x70, Merc. 4EuF	5	(4 MG)	-	1915	1 prototype, built by Union for DMG; crashed Sept. 1, 1915
R.II	4x110, Merc. Fh 1256	3		-	1916/17	1 (450/15), last trials in April 1917
R.I	4x110, Merc. Fh 1256	3			1917	1 (478/15), last trials in May 1917
G.I	?				1917	1 (476/15), not completed
G.II (tractor/pusher)	2x220, Merc. D.IV	3			1917	1 (480/15), 6 were ordered
G.III	2x260, Merc. D.IVa	3			1917	1 (584/16), prototype
Daimler L6 (D.I)	185, Merc. D.IIIbm					1 prototype, 20 ordered
Daimler L8 (CL.I)	185, Merc. D.IIIb					Tests not completed
Daimler L9 (D.II)	185, Merc. D.IIIbm					1 prototype, not finished
Daimler L11 (D.III)	185, Merc. D.IIIbm					Tests not completed
Daimler L14	200, Merc. D.IIIbv					1 prototype finished postwar
Fdh G.II(Daim)	2x200, Benz Bz.IV		2-3 MG, 300 kg bombs			9
Fdh G.III(Daim)	2x260, Merc. D.IVa	3	2-3 MG, 500 kg bombs			~100
Fdh G.IIIa(Daim)	2x260, Merc. D.IVa	3	2-3 MG, 400-500 kg bombs			~180
Fdh G.IV(Daim)						

permanently. If the number was still 500 workers in May 1917, it had grown to 1,500 by the end of the war.

Freiherr von Thüna[2] was appointed director of the Aircraft Construction Department, engineer Schopper initially worked as designers, and from April 1918, Regierungsbaumeister Klemm[3] was appointed chief designer.

Aircraft Development (since the Outbreak of World War I):
An aircraft factory with its own company airfield was built very close to the Böblingen military airfield. Daimler Motoren AG did not play small when it came to the hangars and hangars. However, due to a lack of in-house expertise and skilled personnel, the development contract for the first multi-engine large combat aircraft was awarded to the Union company in Teltow near Berlin. Not such a good choice, because the four-engine Union-Daimler G.I (also R.I) broke down early on, prompting DMG to pursue in-house development under the direction of Karl Schopper, until then the engineer in charge at Union-Flugzeugwerke.

In the spring of 1916, a four-engine large combat aircraft (R.I) was produced, each with two engines in tandem. However, the expectations for this aircraft were not met. This was followed by a lighter aircraft (R.II), equipped with 2 x 220 hp Mercedes engines, which again did not show positive flight characteristics. A third large combat aircraft (G.III) was then built, which had two 260 hp engines in the fuselage. The propellers were driven by a gearbox. The excessive weight of this third prototype was again the reason for insufficient results, so that this type was also not pursued further.

After disappointing results with its own large aircraft developments, Daimler switched to building twin-engine bombers under license from Friedrichshafen. Engineer Schopper was also responsible for the few Daimler fighter

Development of Aircraft Factories

planes: all of them biplanes that did not particularly stand out from the offerings of the effectively operating competitors Fokker, Albatros and Pfalz.

In November 1916, the company signed a license agreement with Flugzeugbau Friedrichshafen and took over the construction of the G-planes. Initially, aircraft of the type Fdh G.II were produced.

In May 1917, the construction of Fdh G.III aircraft was started. Furthermore, the company started manufacturing its own fighter planes.

An interesting personnel change took place on April 1, 1918, when Hanns Klemm took over the management of the design office in Sindelfingen. Under his direction, further prototype fighters were developed shortly before the end of the war and after the capitulation, such as the L.8, L.11 and L.14 types.

Production Supervisors (Bauaufsicht): #15
Production Supervision No. 15 was responsible for the aircraft construction department of Daimler-Motoren-Gesellschaft. This 10-man supervision and acceptance group of the Inspektion der Fliegertruppe (Idflieg) was established at the company on December 25, 1916.

War Material Found by the Inter-Allied Commission of Control (IACC) During Inspections after the End of the War:
79 aircraft
84 engines

Situation after WWI:
Already during the war, the Daimler plant in Sindelfingen produced all kinds of vehicle chassis in addition to engines and airplanes. In 1919, it expanded this production (trucks, passenger cars), tried its hand at making furniture, and began repairing automobiles.

Endnotes
1. Gottlieb Wilhelm Daimler, born Gottlieb Däumler, (* March 17, 1834 in Schorndorf; † March 6, 1900 in Cannstatt near Stuttgart) was a German engineer, designer and entrepreneur. Together with Wilhelm Maybach, Daimler developed the first high-speed gasoline engine and the first four-wheeled motor vehicle with an internal combustion engine.
2. Rudolf Freiherr von Thüna (* April 9, 1887 in Erfurt; † June 9, 1936 in Freiburg im Breisgau) was a German aviation pioneer and entrepreneur. Until 1913, von Thüna served as a lieutenant in the Imperial Army. He was one of the first 20 trainees taught by Simon Brunnhuber at the Döberitz Flying School. On January 27, 1913, he was the first to receive a military pilot's certificate there on the basis of his flying achievements.
3. Hanns Klemm (* April 4, 1885 in Stuttgart; † April 30, 1961 in Fischbachau) was a German engineer and entrepreneur. He was one of the best-known German aircraft designers until late into the postwar period (L15, L20, ...).

Above: Prototype of Daimler L8 biplane, designed das CL.I (1918).

Above: Daimler L6 (D.I) Fighter, prototype. 1917/18.

Above: Daimler L14 was an improved L11, but not finished before the armistice in 1918.

Above: Daimler R.I (1917)

Daimler

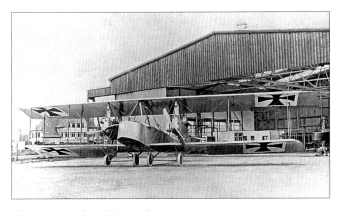

Above: Daimler G.I (1917).

Above: Daimler G.II (tractor) (1917).

Above: Daimler G.II (pusher) (1917).

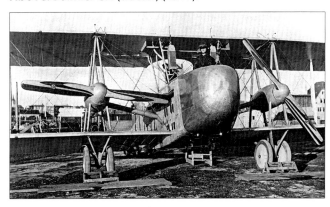

Above: Daimler G.III (1917).

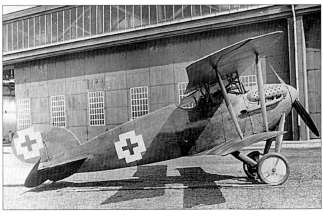

Above: Daimler L6 (D.I) (1917/18)

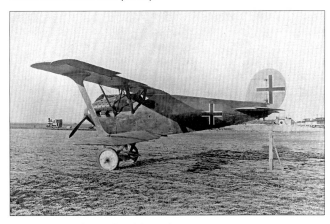

Above: Daimler L9 (D.II) (1918)

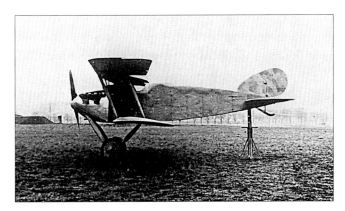

Above: Daimler L8 (CL.I) (1918)

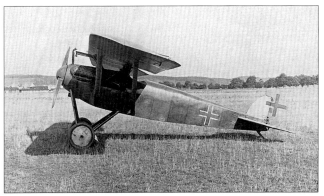

Above: Daimler L11 (D.III) (1918)

Above: Fdh G.II(Daim) (1916)

Above: Fdh G.III(Daim)

Above: Fdh G.IIIa(Daim)

Above: Fdh G.IIIa(Daim) postwar passenger converstion.

Above: Fdh G.IV(Daim)

Above: Daimler L14 in postwar Daimler insignia.

4.8 Deutsche Flugzeugwerke GmbH, Leipzig (DFW)

Foundation:
The company was founded on March 18, 1911, by Kommerzienrat Bernhard Meyer and engineer Erich Thiele[1] at the parade ground in Lindenthal near Leipzig, initially as "Sächsische Flugzeug-Werke" (Saxon Aircraft Works) and renamed "Deutsche Flugzeugwerke GmbH" (German Aircraft Works) as early as November of the same year. Thiele became technical director of the aircraft works. In the same year, he opened the DFW flying school. He succeeded in recruiting pilots Franz Büchner, Eugen Wienczires and Heinrich Oelerich as flight instructors.

In 1914, the Flugzeugwerft Lübeck-Travemünde GmbH was established on the west coast of the Baltic Sea as a subsidiary of Deutsche Flugzeug-Werke Leipzig. See also Chapter 4.22).

In 1917, a second plant was added to the mother company in Leipzig-Großzschocher (Plant II). The subsidiary "National-Flugzeugwerke", originally founded in Johannisthal, was also located on the same terrain.

Factory Facility and Airfield:
At the end of the war, the Deutsche Flugzeugwerke consisted of two separate plants, the Plant I in Leipzig-Lindenthal (northwest of Leipzig) and the Plant II in Großzschocher near Leipzig (southwest), which were merged commercially and technically into a central administration that had its offices in the center of Leipzig (Bosestraße 2).

Plant I Lindenthal covered about 1,073,000 m² of ground area; of this, the airfield accounted for about 1 million square meters, with part of the parade ground included here. The factory site accounted for about 24,000 m², office buildings and casino for about 1,500 m². The D.F.W. flying school is attached to this plant.

The first solid buildings were erected in 1911, and from then on, they were constantly enlarged by new buildings as needed. The solid buildings could not keep up with the development of the plant, especially during the war, for which reason temporary wooden buildings had to be erected, especially to park finished aircraft. In order to secure the factory facilities, the company had established its own fire department, which was responsible for guarding the entire facility with a limited daytime service and an extended nighttime service.

Plant II Leipzig-Großzschocher covered an area of about 1 million square meters, of which 880,000 m² was the airfield and

Above: Map of the German Reich and Leipzig.

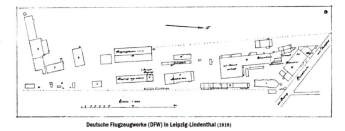

Deutsche Flugzeugwerke (DFW) in Leipzig-Lindenthal (1919)

Above: Site plan DFW in Leipzig-Lindenthal.

Development of Aircraft Factories

Left: Aerial view of DFW plant II in Leipzig Großzschocher.

120,000 m² the factory site. Construction of the new Plant II began in the summer of 1917, and shortly thereafter part of the production was already transferred there from Lindenthal. In addition to the rail siding, which Plant I in Lindenthal did not have, a tramway also ran close to the factory site.

Fabrication:

The following summary gives a further overview of the aircraft produced:
- 1912   2 Mars-Biplane,
- 1913   18 Mars- Biplane,
- 1914   93 Steel-Tauben and „MD 14",
- 1915   79 DFW C.I
- 1916   99 DFW C.I and Dfw C.V
- 1917   820 Dfw C.V and 1 Riesenflugzeug (giants).
- 1918 (til 30. September)   1100 DFW C.V and 2 Riesenflugzeuge (giants).

The ILÜK[2] presented the following production figures in its report published in 1920.

B-type aircraft	99
C- -„-	2,310
R- -„-	6
Total aircraft:	2,445

Staff and Designers:

The personnel grew steadily since the company was founded in 1911 and developed as follows:
Since the foundation of the company the following designers have been working there:
Heinrich Bier and Ingenieur Bourcart (the latter constructed the Mars-Taube and M.D.14) as technical director, Engineer Sabersky[3], Mid 1916 departed (co-designed „C.V"), Director Oelerich[4] (designed the DFW C.V), Engineer Dorner[5] (worked on the R-planes).

Site plan DFW plant II Großzschocher

Since March 1917, the management of the plants was in the hands of the general director Kurt Herrmann.

Part of the workers, about 400 people, went on strike on January 31, 1918 (as in many other armament factories), on the basis of which the Deputy General Command of the XIX Army Corps ordered the militarization of the D.F.W. factories. The military commander became the then head of the construction supervision, First Lieutenant Bonde.

Aircraft Technical Development:

When it was founded in 1911, Deutsche Flugzeugwerke was involved in the construction of various types of aircraft and produced the first useful type, the so-called Mars biplane. In parallel with these experiments, they designed and built the DFW steel Taube (1913/14) and, turning again

DFW

Aircraft Built by Deutsche Flugzeugwerke GmbH, Leipzig (DFW)						
Type (All Army Aircraft)	Engine [hp], Manufacturer	Crew	Armament	License Built By	Year	Notes
DFW Mars Biplane			-		1913	
DFW B.I (MD 14)	100-150, Argus, Mercedes, Benz	2	-,		1914	also with MG
DFW B.II (MRD 15)	100, Merc. D.I	2	1 MG		1915	
DFW C.I (KD15)	150, Benz Bz.III	2	1 MG		1915	40
DFW C.II	150, Benz Bz.III 160, Merc. D.III	2	2 MG		1915	12
DFW C.III	150, Benz Bz.III	2	1 MG		1915	prototype
DFW C.IV (T25)	150, Benz Bz.III	2	1 MG		1916	>5
DFW C.V	200, Benz Bz.IV	2	2 MG	LVG: 400 Av: 1400 Halb: 150	1916	1,000 (600 to front)
DFW C.Vc	185, NAG C.III					
DFW C.VI	220, Benz Bz.IVa	2	2 MG		1917	prototype
DFW C.VII (F37)	200, Benz Bz.IV	2	2 MG		1918	prototype
DFW F 37/III	320, BMW Bmw IV	2	-		1918	Prototype; C.VII for high altitude flights
DFW Flow	100, Merc. D.I	1	1 MG		1916	prototype
DFW D.I (T28)	160, Merc. D.III	1	2 MG		1917	prototype
DFW Dr.I (T34/II)	160, Merc. D.III	1	2 MG		1917	prototype
DFW D.II (F34)	170, Merc. D.IIIa	1	2 MG		1918	prototype
DFW R.I (T26)	4x220, Merc. D.IV	5			1916	R11/15
DFW R.II (T26/II)	4x260, Merc. D.IVa	5	Several MG, 1,700 – 2,600 kg bombs		1918	3 (R15/16 – R20/16)
DFW R.III	8x260 + 1x110		8 MG, 2.500 kg bombs		1918	Project
Halb C.V(DFW)	200, Benz Bz IVü	2	1-2 MG, camera, FT		1915	200
Alb C.III(DFW)	150, Benz Bz.III	2	2 MG		1917	200

Left: DFW advertisement.

Above: Prinz Heinrich at the exhibition stand of DFW (ALA, Berlin 1912).

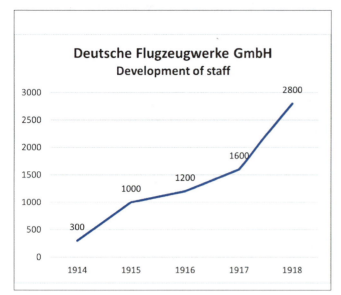
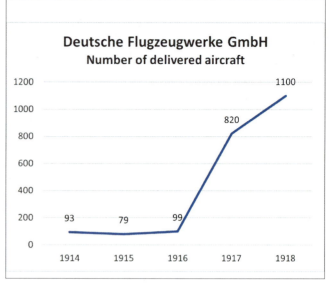

to the biplane, developed a military biplane in 1914 under the designation MD 14. This was adopted by the Army Administration after various required improvements and was then designated Dfw B.I. Further designs on this type then led to a larger order to build Dfw C.V aircraft, a large number of which were produced under license by other German manufacturers. It is also worth mentioning that at the outbreak of the war, the company also received an order for DFW steel pigeon.

Production supervisors (Bauaufsicht): #8

In addition to carrying out its actual task, the Bauaufsicht (Construction Supervision Department as military oversight) established an extensive warehouse in 1917 and took into its own administration all items supplied by the Army Administration to the aircraft company, such as: Engines, radiators, compasses and instruments taken into its own administration.

War Material Found by the Inter-Allied Military Commission of Control During Inspections after the End of the War:

In Lindenthal: 37 aircraft & 25 engines
In Großzschocher: 72 aircraft, including 3 giant R.II aircraft & 41 engines

Situation of the Company after the End of the War:

The plant in Großzschocher was divided. One part went to the transport company Allgemeine Transport Anlagen. One hall was sold to an engine factory. The factory in Lindenthal was sold to an automobile manufacturer.

Endnotes

1 Erich Thiele (* January 2, 1884 in Halle (Saale); † January 13, 1929 in Leipzig) was a German engineer and aviation pioneer and one of the first holders of an official pilot's license for powered flight in Germany. He became known to the Leipzig publisher Bernhard Meyer (1860-1917), who provided him with financial support for aircraft development from 1909. On July 6, 1910 (according to other information on July 10, 1910), Thiele passed the officially prescribed internationally valid pilot examination at the Griesheim airfield in an Euler biplane. He was subsequently awarded German pilot's license No. 13. His flight instructor was the holder of the first German pilot's license, August Euler.

2 ILÜK (Interalliierte Überwachungskommission) - IMCC (Inter-Allied Military Commission of Control).

3 With the transfer of Willi Sabersky-Müssigbrodt from the liquidated Gustav Otto Flugmaschinenwerke in Munich, new ideas came to DFW's design office, where Heinrich Oelerich and Hermann Dorner had already designed the successful DFW B.I and B.II and subsequently, on the same basis, some less spectacular C-planes.

4 Heinrich Bernhard Oelerich (* February 5, 1877 in Hamme; † March 23, 1953 in Freising) was a German aircraft designer and automobile racer. He then joined the Deutsche Flugzeug-Werke (DFW) in Lindenthal as chief pilot. After receiving his German pilot's certificate No. 37 on a Schultze-Herfort monoplane at Berlin-Johannisthal on October 21, 1910, Oelerich undertook demonstration flights in England, France, Portugal and South America. Here he set several records, including a 2 h 41 min endurance flight with two passengers on July 5, 1912, a 6 h 8 min endurance flight on July 8, 1913, and an 8000 m altitude world record on July 14, 1914.

During World War I, Oelerich was a factory pilot, designer and technical director in the DFW, was involved with Hermann Dorner in the design of the DFW B-types and the DFW C.V, and participated in the work on the development of the giant aircraft. In 1918 he crashed while testing the four-engine R.XV and from then on was severely disabled.

5 Hermann Dorner (* May 27, 1882 in Wittenberg; † February 6, 1963 in Hannover) was a German engineer. He was among the first German powered aircraft designers. In 1907, he had begun developing a glider to which an engine could be attached to the nose. In September 1909, he was the only German to participate in the 1st International Flight Week in Johannisthal. He had entered the competition with his self-designed monoplane. On

DFW

July 11, 1910, he won the third Lanz Prize of the Air, worth 3,000 marks, with his T II monoplane, and in August he won another prize at the Johannisthal air show. With the money from the prizes, he opened his Dorner Flugzeug GmbH in 1910. After 1912, however, he was at his wits' end financially and went to work as a flight instructor at the Adlershof Aviation School. In 1913, he went to the Deutsche Versuchsanstalt für Luftfahrt (DVL) - German Research Institute for Aviation, which had been founded shortly before, as technical director. In mid-1914, he developed the DFW B.I and B.II at Deutsche Flugzeug-Werke (DFW) with Heinrich Oelerich. At the end of 1915, he became chief designer of DFW's Giant Aircraft Department in Lindenthal near Leipzig, and from 1916 chief designer for aircraft engine development at Hannoversche Waggonfabrik (Hawa), aircraft construction department, in Hannover-Linden, where he designed the Hannover CL types.

Above: DFW R.II; 3 were completed during the war.

Above: Prewar DFW Arrow biplane.

Above: DFW B.I (MD 14) (1914)

Above: DFW B.II (MRD 15) (1915)

Above: DFW Mars biplane. (1913)

Above: DFW Mars monoplane. (1913)

Above: Armed DFW MD14 (B.I) as predecessor of DFW C-type aircraft.

Above: DFW C.I (KD 15) (1915)

Development of Aircraft Factories

Above: DFW C.II (1915)

Above: DFW C.III (1915)

Above: DFW C.IV (T25) (1916)

Above: DFW C.V (1916/17; most produced German WWI warplane)

Above: DFW C.Vc with C.III Nagb engine (1917)

Above: DFW C.VI (1916/17)

Above: DFW.VII (F37) (1918)

Above: DFW.VII (F37) (1918)

Above: DFW "Floh" (1916)

Above: DFW D.I (T34/1) (1917)

Above: DFW D.II (F34) (1918)

Above: DFW Dr.I (T34/II) (1917)

Above: DFW R.I (1916)

Above: DFW R.II (1918)

Above: Alb. C.III(DFW) (1917)

Above: Halberstadt C.V(DFW) (1918)

Below & Below Right: Serial production at DFW Leipzig.

4.9 August Euler Flugzeugwerke, Darmstadt and Frankfurt a.M. (Eul)

Foundation:

August Euler initially founded his own automotive parts company in 1903. By founding his "Handelshaus für Automobilkonstruktionsmaterial", he successfully monopolized the sale of products from German, French, English and Belgian companies. From 1903 to 1908, Euler had great success selling ignition systems made by Bosch in Stuttgart, which helped him to become wealthy in a short time.

The passion for the automobile was decisive for Euler's further development. The constant exposure to the rapidly developing automotive technology and the associated financial success formed the basis for his real life's work, aviation. August Euler was not only an industrialist, but remained a sportsman, now, however, no longer as a bicycle racer, but as an automobile driver. Through his active participation in various automobile races, including in France, where he won the long-distance automobile race Barcelona-Paris in 1906, contact was made with French aviation pioneers who were also initially addicted to automobile racing. Among them were the future pilots Blériot, Delagrange and Farman.

In 1908, Euler visited the aeronautical exhibitions in Paris. During this visit, August Euler and the Voisin brothers agreed on the basic elements of a license agreement, which included "the exclusive right of sale and the exclusive right of reproduction" of the Voisin apparatuses, including applicable patents in Germany. The contract, which was concluded on January 18, 1908, provided for the delivery of an airworthy apparatus of the latest design by mid-December 1908.

In anticipation of prompt delivery, August Euler intended to set up his factory at his home in Frankfurt am Main. Without being able to designate a specific location for his new company, Euler registered a trade for a flying machine factory with the city of Frankfurt in October 1908. The foundation stone for the oldest German aircraft factory had been laid. The city administration itself had the greatest reservations and was rather "outweighed by the fear of being put in the wrong light with this new-fangled business idea."

Above: Rollout of an early Euler biplane in Griesheim/Darmstadt.

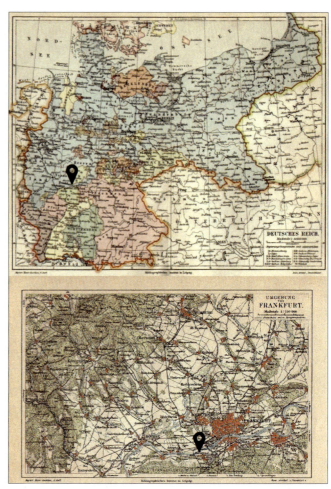

Above: Map Germany and Frankfurt

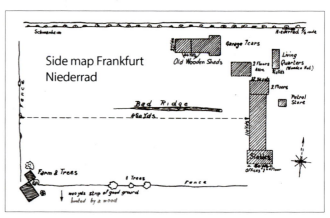

Factory Facility and Airfield:

Since no suitable terrain could be made available to him at first, Euler moved to Darmstadt to the Griesheimer Sand military exercise area, where he remained for 3 years before moving from Griesheim to Frankfurt-Niederrad in January 1912.

It is quite remarkable that at the beginning of 1912 Euler's pilots

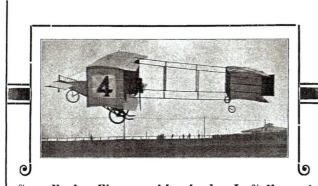

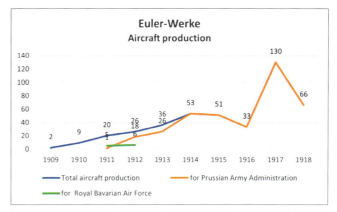

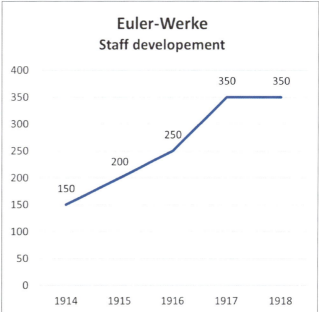

Euler works in Niederrad near Frankfurt.

Above: Typical Euler biplane trainer until 1913/14.

succeeded in flying all 6 aircraft available in Darmstadt by air to the airfield in Frankfurt-Niederrad, about 30 km away, without any failures.

Fabrication:

The Euler Flugzeugwerke didn't play a big role inside the German aviation industry for different reasons.

The following table shows which types were produced for the Prussian Army Administration during the war:

As a sole proprietor without bank capital, August Euler was always a difficult negotiating partner for the military administration.

Development of Aircraft Factories

Aircraft Built by Euler Flugzeugwerke, Darmstadt and Frankfurt a.M. (Eul)

Type (All Army Aircraft)	Engine [hp], Manufacturer	Crew	Arma-ment	License Built By	Year	Notes
Eul B.I	100, Merc. D.I 110, Benz Bz.II	2	-		1912	100 trainers
Lvg B.III(Eul)	120, Merc. D.II	-„-	-„-		1918	43 finished
Lvg C.I(Eul)	160, Merc. D.III	-„-	2 MG		1916	29
Eul D.I	100, Oberursel UI	1	1 MG		1916/17	33 (100)
trainers						
Eul D.II	100, Oberursel UI	-„-	1 MG		1917/18	158

Euler's early aviation successes, e.g. he owned the German pilot's license No. 1, and the first aircraft deliveries to the Bavarian Air Force beginning in 1912, did not continue in the following years. For too long, Euler insisted on his outdated designs with pusher propellers. Euler was initially skeptical of innovations.

The total number of built aircraft over the years did not exceed 395 pieces. around 360 delivered military aircraft were registered in Eulers business books.

August Euler's flight school in Darmstadt-Griesheim trained the first pilots for the Bavarian Air Force. This led to the Bavarian military administration deciding to initially purchase Euler biplanes (Farman type). A total of 11 aircraft were delivered to Bavaria in 1911/12.

Name	Pilot license No., date
Ellery v. Gorissen	4, 21 March 1910
Thiele, Erich	13, 6 July 1910
Lochner, Erich	15, 15 July 1910
Reichardt, Otto	55, 3 February 1911

Staff and Designers:
No well-known designers have become known.

Aircraft Development:
Euler-Werke, which played a minor role in terms of the total number of units produced, had manufactured about 450 aircraft of various types since its founding, delivering 366 of them to the German Army Administration and 11 to the Royal Bavarian Air Force. Until the beginning of the war, these were the company's own designs. During the war, however, only LVG types were built under license.

In the last years of the war, Euler-Werke participated unsuccessfully in the development of several fighter aircraft. Only the D.I and D.II aircraft of their own design (based on Nieuport 11 and 17 respectively) found their way into the air force, mostly as training aircraft.

It should be mentioned at this point that August Euler filed a patent for the installation of a permanently installed machine gun in an aircraft with the patent office in 1910[1]. This patent was offered to the military administration for use, but they refused because of the too high license fees and follow-up contracts. Nevertheless, this patent was used a few years later, which meant that A. Euler was able to assert his legal claims after the end of the war.

War Material Found by the Inter-Allied Commission of Control (IACC) During Inspections after the End of the War:
The airfield at Niederrad near Frankfurt was in French-occupied territory. At the time of the capitulation, Euler-Werke was working on the completion of Lvg B.III(Eul) airframes. All unfinished or unaccepted airframes were moved to unoccupied territory, exactly where is not known. In any case, the report of the Interallied Control Commission for Aviation does not list any war materials found.

Situation after WWI:
The factory served as a cavalry depot for French troops during the occupation of Frankfurt. It was vacated and closed at the end of 1920.

Euler, by then head of the first German aviation authority (Reichsluftamt), sold the Eulerwerke. The airfield site was plowed up and redesigned for an area of allotments.

Endnotes
1 Reichspatent No. DRP 248 601.

Above: Mercedes D.I installation in Eul. B.I.

Euler

Above: Euler-Taube (examplewise). (1913).

Above: Euler flight machine "Yellow dog" in different sizes. (1912–14)

Left: Typical Euler trainer for Prussia (1912–14).

Above: Eul B.I (first batch) with nose wheel. (1914)

Above: Eul B.I (last batch). (1914)

Above: LVG B.III(Eul) with improved center of gravity compared to LVG B.III. (1918)

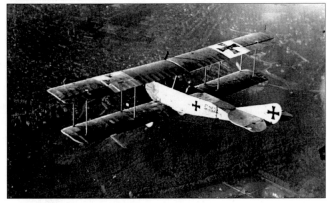

Above: LVG C.I(Eul). (1916)

Above: Eul D.I. (1916/17)

Above: Eul D.II. (1917/18)

Above: Eul two-seater combat airplane (1915)

Above: Eul two-seater combat airplane (1915)

Above: Eul single-seat combat airplane (1915)

Above: Eul D.R.4 prototype (1915)

Euler

Above: Eul triplane trainer prototype. (1915)

Above: Eul D 5 fighter prototype. (1918)

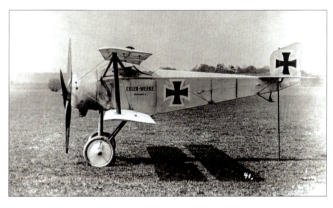

Above: Eul D 6 fighter prototype. (1918)

Above: Eul D.R. 5 triplane fighter prototype. (1917/18)

Above: Eul D.R. 7 triplane fighter prototype. (1917/18)

Above: Eul D.R. 9 triplane fighter prototype. (1917/18)

Above: Eul D 4 quadraplane fighter prototype. (1917)

4.10 Fokker Flugzeugwerke, Schwerin i. Mecklenburg (Fok)

Above: Advertisement 1915.)

Foundation:
Fokker Flugzeugwerke was founded by the Dutch engineer Anthony Fokker[1] on October 1, 1913 as a limited liability company. He came to Germany at the beginning of 1912 to settle in Johannisthal, where he first founded the "Fokker Aviatik GmbH". In the meantime, other company names were used, such as "A.H.G. Fokker Aeroplanbau" and "Fokker Aeroplanbau GmbH". After the success in Johannisthal with the construction of the Fokker Spider did not materialize, Fokker moved his company to Schwerin in Mecklenburg, where he opened the company "Fokker Flugzeugwerke" on October 1, 1913. The abbreviated designation assigned by the Flugzeugmeisterei was "Fok".

The Fokker aircraft factories were merged with Junkers in 1917 at the insistence of the military administration to form "Junkers-Fokker Flugzeugwerke A. G.".

Factory Facility and Airfield:
In 1913, Anthony Fokker purchased two parcels of land from the city of Schwerin, on each of which the following buildings were erected:
1) The Fokker factory;

Above: Map of Germany and Schwerin.

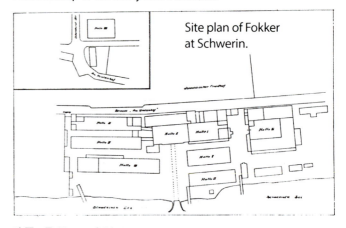

2) The Fokker airfield.
The buildings in Schwerin consisted mainly of barracks. The

Fokker

> Wegen Vergrösserung unseres gesamten Betriebes verlegen wir am 1. Oktober 1913 unsere Firma nach
>
> **Schwerin i. Meckl.**
>
> Eigener Flugplatz u. Wasser-Flugplatz
>
> sowie moderne und grosse Fabrikräume stehen uns nunmehr zur Verfügung, um auch grössere Aufträge in kürzester Zeit ausführen zu können.
>
> **FOKKER-Aeroplanbau**
> m. b. H.

Above: Fokker Aeroplanbau advertises their new location in Schwerin in 1913 after moving from Johannisthal.

50-hectare airfield was located 35 km away from Schwerin near the village of Görries on the Schwerin-Berlin railroad line.

New barracks and halls were constantly being built next to the factory and on the airfield. In Schwerin, he acquired the Perzina piano factory in order to have wood specialists at his disposal. Other branch factories were added, for example the piano factory Adolf Nützmann in Schwerin, the Zimmermann arms factory in Berlin-Reinickendorf and Magyar Általános Gépgyár in Hungary; in 1916 he acquired a sixth of the shares in Motorenfabrik Oberursel AG, whose rotary engines were used to equip a large number of his fighter planes; in October 1917 he became co-owner of Flugzeugwerft Lübeck-Travemünde GmbH.

Fabrication:
The Fokker factory in Schwerin produced a total of at least 3,350 aircraft of various types, 150 of which were assembled after December 11, 1918, and in 1919.

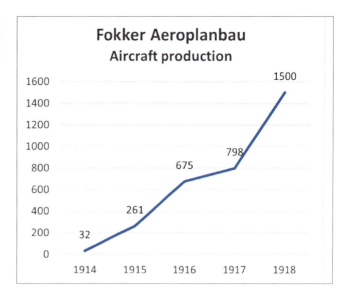

Fokker factory was specialized in building fighter planes. Its first planes were copies of the "Morane". The "E.I." was the first German aircraft equipped with a machine gun that fired through the propeller. Shooting through the propeller contributed much to the success of Fokker aircraft. In particular, their D aircraft played a prominent role in 1916 and again from early 1918 (D.VII).

Staff and Designers:
The number of Fokker's employees increased during the war from about 100 to about 6000, of which he employed over 1200 in Schwerin alone.

Aircraft Development:
The company was engaged, with one exception, exclusively in the manufacture of aircraft of its own design. The following were produced: Type E.I–IV (with fixed MG), Type D I–III, Type Dr.I, D.VI, D.VII, and Type E.V/D.VIII.

The mainly handicraft method of production in the Fokker works prevented larger production numbers in the first year of the war, which meant that the proportion of Fokker machines in front-line aircraft was small. In the first quarter of 1915, the front-line inventory lists showed only 29 A and B aircraft, compared with 637 aircraft of these classes in April 1915.

On April 18, 1915, when a hit from the ground forced the well-known French pilot Roland Garros to land behind German lines, the Germans discovered a machine gun permanently installed on the Morane Type L monoplane. Fokker and his engineers created a mechanical engine-driven machine gun synchronization system which, after initial difficulties, ensured that it could be fired through the propeller circle of the aircraft. This was first realized on a Fokker M5K (A.III) one month after Garros' capture and demonstrated to the Traffic Engineering Review Board (V.P.K., VauPeKa).

The appearance of the Allies' light, maneuverable biplanes, also equipped with synchronized MG, heralded the end of the Fokker monoplane.

Aircraft Built by Fokker Flugzeugwerke, Schwerin i. Mecklenburg (Fok)

Type (All Army Aircraft)	Engine [hp], Manufacturer	Crew	Armament	License Built By	Year	Notes
Fok A.I (M.8)	80, Oberursel U.0	2	-	Halb (A.II)	1914	80
Fok A.II (M5L)	80, Oberursel U.0	2	-„-		1915	18 +4
Fok A.III (M5K)	80, Oberursel U.0	1	1 MG		1915	5
Fok B.I (M7)	80, Oberursel U.0	2	1 MG		1915	~20
Fok B.I (M10E)	80, Oberursel U.0	2	-		1915	~55
Fok B.II (M10Z)	100, Oberursel U.I	2	-		1915	15
Fok C.I	185, BMW Bmw.IIIa	2	2 MG		1918	few
Fok E.I (M5KMG)	80, Oberursel U.0	1	1 MG		1915	48
Fok E.I (M14)	80, Oberursel U.0	1	1 MG		1915	
Fok E.II (M14)	100, Oberursel U.I	1	1 MG		1915	49
Fok E.III (M14)	100, Oberursel U.I	1	1 MG		1915	~270
Fok E.IV (M15)	160, Oberursel U.III	1	2 MG		1915	50
Fok E.V (D.VIII)	110, Oberursel Ur.II 160, Oberursel Ur.III	1	2 MG		1918	289 (335)
Fok D.I (M18)	120, Merc. D.II	1	1 MG		1916	113
Fok D.II (M17)	100, Oberursel U.I	1	1 MG		1916	181 +22
Fok D.III (M19)	160, Oberursel U.III	1	2 MG	MAG	1916	210
Fok D.IV (M21)	160, Merc. D.III	1	2 MG	MAG	1916	40
Fok D.V (M22)	100, Oberursel U.I	1	1 MG		1916/17	300
Fok D.VI	110, Oberursel Ur.II	1	2 MG		1918	60
Fok D.VII	160, Merc. D.III 170, Merc. D.IIIa 180, Merc. D.IIIaü 200, Merc. D.IIIaüv 185, BMW BMW IIIa 185, Mana III	1	2 MG	Alb: (1000) OAW: (1300)	1918	(~1000) ~2500 all
Fok DR.I	110, Oberursel Ur.II	1	2 MG		1917	320
Fok F.I	120, Oberursel Ur.II	1	2 MG		1917	3
AEG C.IV(a)(Fok)	160, Merc D.III	2	2 MG		1917/18	300 (400)
AEG C.IVa(Fok)	180, Argus As.III	2	2 MG			

Fokker had produced four models, the E.I with 80 hp was basically an MG-equipped M5K. The internal factory designation M14 was given to the E.II and E.III with 100 hp rotary engine U.I. The last variant to appear in October 1915 was the E.IV (M15) with a 160 hp Oberursel U.III rotary engine. This monoplane could carry 2 or even 3 MG, but by this time it was already out of date.

From mid-1916, the faster and more maneuverable biplanes displaced the E-planes from service. Of this new category, the D airplanes, Fokker brought out the models D.I to D.IV from mid-1916. In practice, however, it became apparent that the Albatros machines were more powerful, which led to the Fokker D-planes soon being withdrawn from the front.

In January and June 1917, Fokker-Flugzeugwerke received an order from the Inspektion der Fliegertruppen (Idflieg) to build 200 Aeg C.IV reconnaissance planes under license, since these had a steel fuselage construction similar to the company's own machines. This fact, considered an embarrassment for A. Fokker, changed only after the appearance of the first English Sopwith triplanes. Fighter aces, such as Manfred von Richthofen and others, demanded equal aircraft, which led to a demand for agile fighter triplanes from the German aircraft industry. Fokker offered the first design as early as July 1917. Two prototypes 102/17 and 103/17 (factory designations D.VI, militarily declared

Fokker Aircraft Performance Specifications

Type	Empty Weight (kg)	Max Speed (km/h)	Time to climb to Altitude in Minutes					
			1000m	2000m	3000m	4000m	5000m	6000m
E.I	358		7	20	40			
E.III	399	130	5	15	30			
E.IV	466	140	3	8	15	25		
D.I	463	160	5	11	16	28		
D.II	384	150	4	8	15	24		
D.III	452	150	3	7	12	20		
D.IV	606	160	3	5	12	20	30	
D.V	363	170				19		
Dr.I	375	200	1¾	3¾	6½	10	14½	
D.VI	393	200	½	5½	9	13½	19	
D.VII	688	200	1¾	4	7	10¼	14	18¾
D.VIII	405	200	2	4½	7½	10¾	15	19½
V 36	637	200	1¾	4	6¾	10	13½	18½

F.I) were sent to Jagdgeschwader 1 for front-line testing, where they were flown by Werner Voss and Richthofen, among others. The Dr.I was to be a success, 20 aircraft *were ordered until, however, the upper wing broke off on two aircraft during* front-line operations. This resulted in all Fokker Dr.I being grounded. After design changes were submitted and approved, series production continued. In the end, 320 Dr.I were delivered.

The Idflieg organized three competitions for D aircraft in Berlin-Adlershof in January 1918. All German aircraft factories were invited to present their aircraft. Fokker did so with the prototypes V11 and V18, each equipped with Mercedes D.III engines. As a result of these comparison flights, the type later designated Fok D.VII was selected for series production. To meet the large demand, license orders were placed with Albatros-Werke in Johannisthal and Ostdeutsche Albatros-Werke (OAW).

Fokker's last design before the end of the war was the Fok D.VIII, again a monoplane, designed as a cantilever high-wing aircraft. 289 fighters of this type were built. It can be assumed that the influence of Hugo Junkers played a role in the construction, because in the meantime there was a (forced) merger with Junkers-Werke from Dessau.

Performance table of selected Fokker aircraft

Production supervisors (Bauaufsicht): #13

The 19-man crew of the Bauaufsicht, which had to be housed and fed by the company, consisted of 3 officers, a commissioning pilot, four plant foremen, and crews to ensure their own duty operations and to support production.

War Material Found by the Inter-Allied Commission of Control (IACC) During Inspections After the End of the War:

442 aircraft,
539 Engines
and several spare parts.

Situation after WWI:

Previously (1919), the complete inventory was transferred to the Netherlands. Here Fokker continued to develop successfully. They were temporarily the only Dutch manufacturer of civil airliners until their insolvency in 1996.

In 1920, the former Fokker Flugzeugwerke GmbH was transformed into "Schweriner Industrie Werke GmbH" and manufactured motorboats, autogenously welded iron beds and sheet metal furnaces. At the end of 1920, it built some civil aircraft of the "Limousine" type with 185 hp BMW engine.

For these various works it employed about 150 workers.

Endnotes

1 Anton Herman Gerard „Anthony" Fokker (* 6. April 1890 in Kediri/Java; † 23. December 1939 in New York) was a Dutch-German-US aircraft manufacturer.

Development of Aircraft Factories

Advertisement 1914.

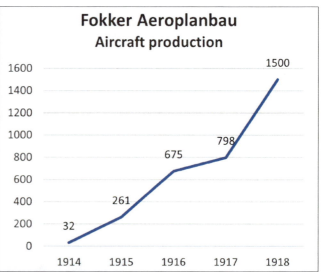

Above: Fokker M1 monoplane (1913).

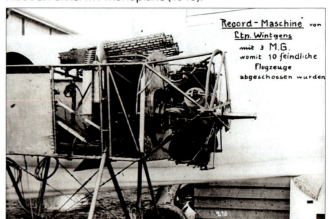

Above: Three guns on Wintgens Fokker E.IV.
Left: Designer and fighter pilot: Fokker and Wintgens.

Above: Anthony Fokker in front of his A.II aircraft.

Fokker

Above: The management of Fokker-Werke celebrates the completion of the 500th Fokker aircraft: a 1-bay Fok D.II.

Above: Fok A.II (M5L) (1915)

Above: Fok A.I (M8) (1914)

Above: Fok A.III (M5K) (1915)

Above: Fok B.I (M7) (1915)

Above: Fok B.I (M10E) (1915)

Above: Fok B.II (M10Z) (1915)

Above: Fok E.I (M5KMG) (1915)

Development of Aircraft Factories

Above: Fok E.II (M14) with experimental cowling. (1915)

Above: Fok E.III (M14) (1915)

Above: Fok E.IV (M15) (1915)

Above: Fok E.V (Fok D.VIII) (1918)

Above: Fok D.I (M18) (1916)

Above: Fok D.I (M18E) (1916)

Above: Fok D.II (M17) (1916)

Above: Fok D.II (M17E single-bay) (1916)

Fokker

Above: Fok D.III (M19) early production with wing warping (1917)

Above: Fok D.III (M19) late production with ailerons. (1917)

Above: Production Fok D.IV (M21) (1916)

Above: Fok D.IV with experimental nose (M21). (1916)

Above: Fok D.V (M22) (1917)

Above: Fok D.VI (1918)

Above: Fok D.VII (1918)

Above: Fok F.I (V5), Richthofen. (1917)

Development of Aircraft Factories

Above: Fok Dr.I (1917)

Above: Fok C.I (1918)

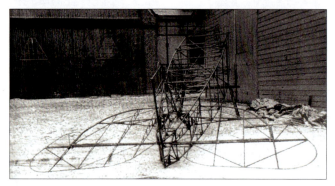

Above: AEG C.IV(Fok) uncovered airframe (1918)

Above: Fok V17, fastest fighter at 1st Fighter Competion (1918)

Above: Fok V20 (1918)

Above: Fok V23 (1918)

Above: Fok V25 (1918)

Above: Fok V27 (1918)

Above: Fok V29, winner of 3rd Fighter Competition. (1918)

Above: Fok V37 (1918)

4.11 Germania Flugzeugwerke GmbH, Leipzig-Mockau

Foundation:
The beginnings of Germania-Flugzeugwerke GmbH date back to 1910, when the former director of the Erwin Leiber company in Göhrwihl in the Black Forest undertook experiments with wooden strip tubes for aircraft fuselages and wing spars. In 1912, the company was founded under the name "Rathjen & Co." in Berlin-Teltow, and on December 17, it was transformed into the above-mentioned company.

Subsequently, Works II (in Leipzig-Mockau, Leipziger Str. 200b) and Works III in Leipzig-Sellerhausen, Torgauer Strasse, were established as subsidiaries.

Factory Facility and Airfield:
The company settled on the premises of "Luftschiffhafen- und Flugplatz-A.G., Leipzig and rented three existing hangars there in December 1914. In August 1915, seven additional hangars were added. In September 1916, Plant II in Leipzig-Mockau was purchased and one month later, in October 1916, the assembly hall was extended. In October 1917, a parking and dismantling hangar for repair aircraft was built and Plant III was purchased. Then, in November 1917, a new carpentry shop II was built, and in December of the same year a new launching hangar for new and repair aircraft was built. The modernization of the production facilities continued steadily until the end of the war.

Germania-Flugzeugwerke had its own extended railway siding, which was located at the 300 x 500 m airfield of Luftschiffhafen- und Flugplatz-A.G. This was useful for the delivery of aircraft to be repaired and new aircraft, as well as for their transport to the front. The supply of the factory with all the necessary material was also handled for the most part via this siding.

Fabrication:
Germania-Flugzeugwerke was originally engaged in the construction of self-designed aircraft, especially for the naval administration and allied armies, and later manufactured new aircraft under license and set up a large repair shop for military aircraft and attached to it a procurement office for aircraft spare parts.

Since 1917, about 310 licensed aircraft were built and repair work was carried out on Ru C.III and Ru C.IV during the war. The number of aircraft repaired reached 12 per month, but this was only 50% of the capacity of the plant. Shortly before the end of the war, the 300th repair machine was delivered back to front.

Staff and Designers:
Erwin Leiber was also the managing director and chief designer. The personnel grew from an initial 78 men (January 1916) to 10 times that number in January 1918. At the end of 1918, the company's staff numbered about 850. Among the pilots were Karl Krieger[1], who was killed in a Germania plane crash, and Otto Onigkeit[2].

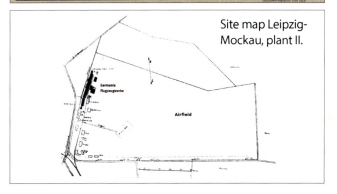

Site map Leipzig-Mockau, plant II.

Development of Aircraft Factories

Aircraft Built by Germania Flugzeugwerke GmbH, Leipzig-Mockau						
Type (All Army Aircraft)	Engine [hp], Manufacturer	Crew	Armament	License Built By	Year	Notes
Germania Taube	100, Argus As.I	2	-		1914	Several a/c
Germ B.I	120, Argus As.II	2	-		1914	2 for Navy (Rathjen)
Germania JM	100, Argus As.I	1	1 MG		1916	1 prototype
Germania DB (Typ D)	180, Argus As.III	2	-„-		1916	1 prototype
Germania KDD	150, Benz Bz.III	2	-„-		1916	1 prototype
Germania C.I	260, Maybach Mb.IVa	2	2 MG		1917	1 prototype
Germania C.II	260, Maybach Mb.IVa	1	1 MG		1917	1 prototype
Germania C.III	260, Maybach Mb.IVa	2	2 MG		1917	1 prototype
Germania C.IV	180, Argus As.III	2	-		1917	1 prototype
Ru C.I(Germ)	150, Benz Bz.III	2	1 MG		1917	100 (40 with Rapp Rp II V8)
Ru C.Ic(Germ)	185, NAG Nag C.III	2	1 MG		1917	200

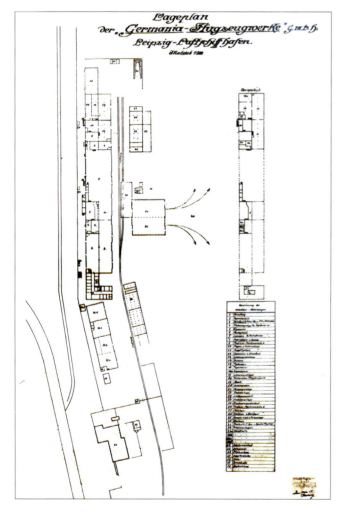

Aircraft Development:

In 1912, a Taube with a round fuselage and with wings made of wooden strip tubes was the first of this design. This design represented a significant advance for the time, as this first "Wickelrumpf" (wounded fuselage) (made of wooden strip) can be considered the forerunner of the later veneer fuselages. With further improvements, two more pigeons of this design were built, equipped with 100 hp Argus engines, which were accepted by the Army Administration and the Navy in 1914 according to the regulations prescribed at that time. The factory supplied mainly aircraft for the Navy administration and somewhat later brought out in parallel a biplane as a B-machine, which also met the acceptance requirements. A C-machine type following this B-machine was also successfully introduced into the Reichsmarine. In addition to the licensed production of mainly Albatros aircraft, the company also developed its own interesting aircraft types, e.g. the Germania KDD (Kampf-Doppeldecker), JM (Jagd-Maschine) or several designs of C-aircraft.

In summary: The development of its own types of aircraft that were suitable for war failed.

In 1915, the first order was placed by Idflieg for aircraft repairs and for the supply of spare parts. New aircraft for the Air Force were ordered for the first time in the autumn of 1917. Germania Flugzeugwerke received an order to supply 100 new Ru C.I. aircraft.

In May 1917, the company received a further order for 100 Ru C.I aircraft, which was followed in October of the same year by another order for 100 aircraft.

Left: Detailed plan view of Germania works.

Germania

The aircraft repair shop had expanded in accordance with the requirements of the Army administration to such an extent that in the fall of 1917 a closed order was placed for 100 Ru C.IV type aircraft for repair.

To eliminate the cooling problems of the 260 hp Maybach Mb.IVa engine, various cooling methods were tried: a) with so-called ear coolers next to the fuselage, b) cooling surfaces on the struts between the wings, or c) using surface coolers.

At the beginning of 1918, at the instigation of the Idflieg, the production figures were initially reduced to 30 new aircraft per month, and later to 25. The number of repairs was also reduced (only up to 12 repairs per month), which meant that only half of the plant's capacity was utilized.

Production Supervisors (Bauaufsicht): #22
Production Supervision No. 22 was last headed by First Lieutenant Schäfer.

War Material Found by the Inter-Allied Commission of Control (IACC) DuringBInspections After the End of the War:
67 aircraft,
93 engines.

Situation after WWI:
At the end of 1920, the former Germania factory, meanwhile split up and renamed "Mitteldeutsche Möbelfabrik Leipzig-Mockau" and ""Werkstätten für Mechanik GmbH", produced iron beds, furniture and agricultural machinery. It had rented a hangar and an office from the airline "Deutsche Luft Reederei" (DLR). Until then, DLR used the hangar as a repair workshop.

Endnotes
1 Nineteen-year-old Prince Sigismund built a flying machine in a workshop he had set up in Alt-Glienicke Castle near Potsdam. This apparatus was tested with good success by Karl Krieger, who had already exhibited a self-built aircraft model at the "ILA" (1909). He later built his own airplane in Johannisthal with imperial support, on which he learned to fly without an instructor and passed his pilot exam after eight weeks. Krieger belonged to the old Johannisthal guard, who preferred to starve and beg for money rather than abandon their passion for flying. Time and again he lacked the means to realize his own design ideas. So he flew in other people's machines, first Kühlstein's racing machines, then the Jeannin monoplane and finally the Prince's plane.

Around 1914, the Krieger brothers designed the very first torpedo aircraft and also filed the corresponding patents. As a result, all components and design documents were confiscated by the Reichsmarineamt. In the subsequent lawsuit brought by the Krieger brothers, the attorneys of the Reichsmarineamt requested imprisonment for both of them for alleged military sabotage, but the Amt was instead sentenced to pay 500,000 gold marks in damages, in return for which it received all the design documents and patents at its free disposal. This sum was not paid out because of the war. The Krieger brothers' business was closed down around 1915/16 and Karl Krieger went to the "Germania-Flugzeugwerke" in Leipzig as an acceptance pilot, where he crashed fatally for unknown reasons on August 30, 1918.

2 Otto Onigkeit, Magdeburg, Leipzig. This old eagle, a painter by profession, had been involved in powered flight since 1909 and had received his pilot's certificate No. 368 in April 1913. He built four monoplanes before the War; the last one, built in 1914. It was powered by a 50 hp RAW engine. During the war, Onigkeit worked for LVG in Köslin as a flight instructor, then briefly for Schwade in Erfurt, and then for Germania-Werke in Leipzig as a flight instructor and acceptance pilot.

Above: Germania works entry gate in Leipzig-Mockau.
Right: Germania advertisement 1913.

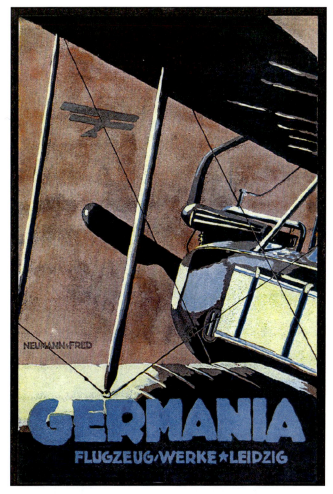

Development of Aircraft Factories

Above: Germania Flugzeugwerke. In front stands a Rumpler C.I(Germ) built under license.

Above: Germania Taube production. Six-cylinder Argus As.I and Wolff propellers have been used.

Above: Advertisement showing different wooden band products made by J. E. Leiber.

Above: Germ C.I with radiators on the struts between the wings.
Right: Germ C.I with ear radiators on both sides of the fuselage.

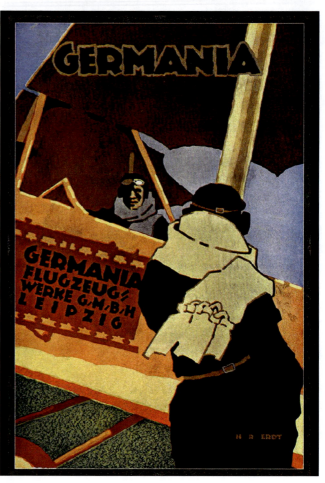

Above: Germania advertisement 1913.

Above: Presentation of the 300th repaired Rumpler aircraft (Ru C.IV).

Germania

Above: Germ C.I with radiators on the cabane struts. (1918)

Right: Mass production of Rumler Ru C.I(Germ). well visible are the Benz Bz.III engines.

Above: Germania Taube. (1914)

Above: Germania B.I (1914)

Above: Germania JM. (1916)

Above: Germania DB (Typ D). (1916)

Above: Germania C.I with airfoil radiators. (1918)
Left: Germania KDD. (1916)

Development of Aircraft Factories

Above: Germania C.II (1917)

Above: Germania C.III (1917)

Above: Rumpler C.I(Germ)

Above: Rumpler C.Ic(Germ)

Above: Germania C.II (1917)

Above: Germania C.IV (1917)

Above: Germania advertisement.

4.13 Gothaer Waggonfabrik A.-G., Gotha (Go)

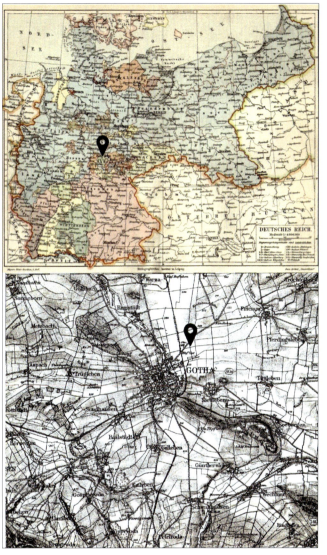

Above: Map of Germany and Gotha.

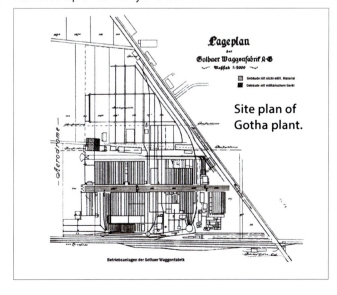

Site plan of Gotha plant.

Foundation:
The company was founded on July 1, 1898 in Gotha as a wagon factory. Subsidiaries in Fürth (Bavaria) and Warnemünde were under construction until the end of the war. However, aircraft construction did not start until the end of 1912.

Factory facility and airfield:
In 1912, when aircraft production began, the Gotha Waggonfabrik comprised two halls with a combined area of 4,220 m² and subsequently expanded production by adding further factory halls, most of which already existed, which were used for the manufacture of railroad cars in peacetime. The development in terms of area was as follows:

The company had its own airfield of about 300,000 m² attached to the factory.

In 1915 and 1916, the "Versuchsbau Gotha Ost GmbH" temporarily operated on the site of the Waggonfabrik (see section on "Zeppelin").

Fabrication:
The company started to manufacture airplanes at the end of 1912 and was engaged in the production of land and sea

Aircraft Built by Gothaer Waggonfabrik A.-G., Gotha (Go)

Type	Engine [hp], Manufacturer	Crew	Armament	License Built By	Year	Notes
Landplanes						
Go A.I (LE1 – LE4)	75 -100, Merc. E4F and D.I	2	-		1913/14	50
LD1, 1a (B-type)	100, OU U.I	2	-		1914	1
LD2 (B-type)	100, Merc. D.I	2	-			
LD3, 4	75, Gnome	2	-			1+1
LD5 (B-type)	100, OU UR.I	2	-			~13
LD6a	150, Benz Bz.III	2	-			1
Go B.I (LD7)	160, Merc D.III	2	-		1914	18
Go B.II (LD10)	100, OU U.I	2	-		1914/15	10
Go G.I	2x150, Benz Bz.III 2x160, Merc. D.III	2-3	2 MG, 250 kg bombs		1915	8
Go G.II	2x220, Merc. D.IV	3	2 MG, 450 kg bombs		1916	10
Go G.III	2x260, Merc. D.IVa	3	3 MG, 500 kg bombs		1917	25
Go G.IV	2x260, Merc. D.IVa	3	3-4 MG, 500 kg bombs	SSW: (80), LVG: (150+40)	1916/17	52
Go G.V	2x260, Merc. D.IVa	3-4	2-3 MG, 350 kg bombs		1916/17	100
Go G.Va	2x260, Merc. D.IVa	3-4	3-4 MG bombs		1917	23
Go G.Vb	2x260, Merc. D.IVa	3-4	3-4 MG bombs		1917/18	80
Go G.VI	2x260, Merc. D.IVa	3-4	3-4 MG bombs		1917/18	2
Go G.VII/GL.VII	2x260, Merc. D.IVa 2x260, Maybach Mb.IVa	3	1 MG bombs	Av: 100 LVG: 30	1918	12
G.VIII/GL.VIII	2x260, Maybach Mb.IVa	2	1–2 MG bombs		1918	prototypes
Go G.IX	2x260, Maybach Mb.IVa	2	1–2 MG bombs	LVG: 70	1918	30?
Go G.X	180, BMW BMW.IIIa	2	1–2 MG bombs		1918	2

planes and their spare parts. The factory was equipped with the most modern woodworking and metalworking machines. The manufactured parts were tested by a well-developed control system. A materials testing center provided strength testing of incoming materials and manufactured structural parts.

The main aircraft components were manufactured in-house, and only a few special parts were purchased from

Gotha

Aircraft Built by Gothaer Waggonfabrik A.-G., Gotha (Go)						
Type	Engine [hp], Manufacturer	Crew	Armament	License Built By	Year	Notes
Seaplanes						
WD1	100, Merc. D.I	2	-		1914	6
WD2	150, Rapp Rp.III 150, Benz Bz.III 160, Merc. D.III	2	1 MG		1915/16	22 for different countries
WD3	160, Merc. D.III	3			1915	Prototype
WD4 (UWD)	2x160, Merc. D.III	3	1 MG, bombs		1916	1
WD5	160, Merc. D.III	2	bombs		1915/16	1
HaBra NW(Go) Gotha designation: WD6	160, Merc. D.III	2	50 kg bombs		1916	30
WD7	2x120, Merc. D.II 2x100, Merc. D.I 2x120, Argus As.II	2	1 MG		1915/16	1 6 1
WD8	240, Maybach Mb.IVa	2	2 MG			8
WD9	160, Merc. D.III	2	1 MG		1916	1
WD10 (U-1)	150, Benz Bz.III	1			1916	1
WD11	2x160, Merc. D.III	3	1 MG 1 Torpedo		1916	17
WD12	160, Merc. D.III 150, Benz Bz.III	2	1 MG, 50 kg bombs		1917	7+14
WD13	150, Benz Bz.III	2	1 MG			18+2 6 (Turkey)
WD 14	2x220, Benz Bz.IV	3	1 MG 1 Torpedo		1917	52 69
WD15	260, Merc. D.IVa	2			1917	2 prototypes
WD20	2x260, Merc. D.IVa	2	1 Torpedo		1917	3 prototypes
WD22	2x160, Merc. D.III + 2x120, Merc. D.I	2			1918	2
WD27	4x175, Merc. D.IIIa	2			1918	prototype

specialized firms.

The Gotha Waggonfabrik built (excluding prototypes):

Landplanes:
- 50 A-aircraft
- 37 B-aircraft
- 305 G-aircraft

Seaplanes:
- 92 single engine seaplanes
- 96 flying boats with 2 engines and
- 2 seaplanes with 4 engines

This is a total of 387 land aircraft and 190 seaplanes, whose various types are listed in the adjacent tables.

Staff and Designers:
The personnel in aircraft construction grew as follows: Engineer Burkhard and Dipl.-Ing. Rößner worked as designers.

Aircraft Development:
Landplanes:
In 1913, construction of the first Taube began, still in a primitive design with a trapezoidal fuselage covered with fabric and fitted with a large tangle of bracing wires. The undercarriage was the original of the Taube designs with

Development of Aircraft Factories

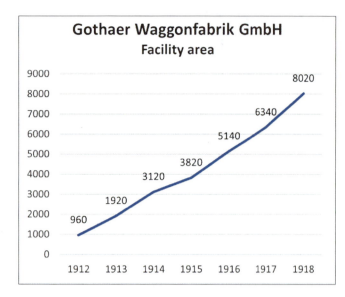

hinged wheels. The first Tauben were equipped with 70 hp 4 cyl; 75 hp 6 yl, 100 hp 6 cyl, Mercedes and 100 hp 4 cyl Argus engines. The climbing ability of the pigeons designed at that time, which had a total weight of about 650 kg and handled a payload of 250 kg, was good. They reached 800 m altitude in 15 minutes, and the speed was between 80 and 95 km/h.

At the end of 1913, trials of light and heavy biplanes were undertaken. Based on the French model, Coudron design, the LD3 light biplane was equipped with a 50 hp Gnome rotary engine with 7 cylinders. The heavy land biplane LD6 was three-strut with plywood fuselage and 150 hp Benz Bz.III engine. With this, also the construction of seaplanes began and the contact with the Imperial Navy, which had good successes with this biplane, mounted on floats.

In 1913, only the Gotha-Taube was produced in series, the improvement of which, especially in the comfortable arrangement of the pilot's seat, resulted in a substantial increase in the dead weight by widening the fuselage. It should also be mentioned that the bracing and controls were also improved, which made the Gotha-Taube popular as a comfortable, safe and easy-to-fly representative of military aviation.

The year 1914 began with the insistent demand for better flight performance of all Tauben in general, which could no longer be postponed. Although the "Tauben" in general showed significant progress in the development of flight technology, they could not compete with the biplanes, which were in fierce competition. The company soon decided to reduce the development work on the Tauben and turned to the production of biplanes in series. The first light biplane, at that time also called cavalry biplane, was produced, as well as a heavy one, which was called B-machine. The light biplane was soon abandoned, however, because the pilots were not happy with the built-in Gnom rotary engine. The company had better luck with the heavy biplane, which was equipped with a 120 hp 6-cylinder Mercedes engine.

The old popularity of the Taube in the field did not allow it to disappear so quickly in production, because a number of pilots who had used the Taube in the beginning could not let go of it so quickly. For this reason, they tried to adapt it better to the requirements of war, developing an enlarged tank, more comfortable interior and easier rigging.

The Tauben had remarkable flight successes on September 2, 1914, when they circled over Paris in a squadron flight. It was reserved for the Gotha Taube to be the first aircraft (pilot Lt. Caspar) to fly to England (Dover) during the war.

It is also worth mentioning that in cross-country flights the Taube already had significant successes in pre-war times. In the national flight competition, Schlegel won the 1st prize of 60,000 marks on the Gotha Taube after Stoeffler's world record, and Lieutenant Caspar won the 2nd prize of 50,000 marks.

At the beginning of 1915, the question of armament and thus the combat value of the flying machine became increasingly important, as front-line flying led more and more to aerial combat. It was out of this need that the so-

Above: Lieutenant (ret.) Caspar achieved a German altitude record on Gotha-Taube. (Engine: V8 Aeolus).

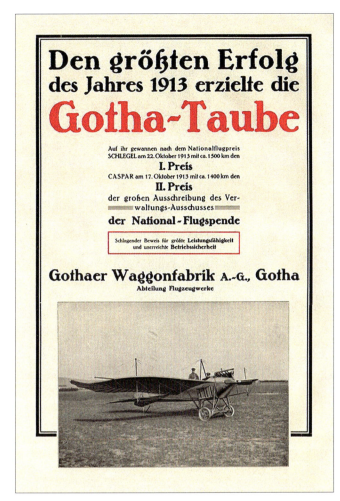

Above: Advertisement 1913.

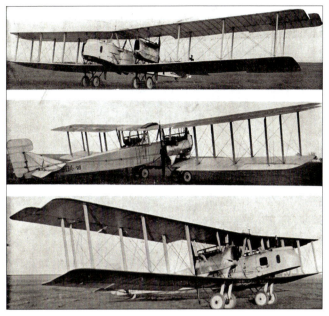

Above: Gotha G.V bomber (1917).

Above: The great successes around the national flight donation. The distance of 2,160 km covered by Viktor Stoeffler corresponds to a distance from Berlin to the center of Portugal or Malaga.

Above: Gotha G.I (Design Ursinus), 1915.

called C-machine was born. The pilot's seat, which used to be behind the observer (especially on aircraft with rear-mounted propellers), was moved to the front, and the aircraft was equipped with a machine gun that could be moved in all directions, followed shortly thereafter by a fixed machine gun that could be operated by the pilot. The C-machines released by the company did not go into series production, as they had already been overtaken by the competition in terms of performance while the test machines were being completed.

In March 1915, Gothaer Waggonfabrik acquired the license of a giant fighter aircraft with 2 x 160 hp Mercedes engines and a high-mounted fuselage, "Friedel Ursinus" type. The company produced about 20 such G-planes in 1915, and at the same time completed the work in progress on B-planes with 120 hp Mercedes engines. Shortly thereafter, the company brought out its own design of a G-airplane, since the Ursinus design could not hold up in the long run due to its heavy armor with a high-mounted fuselage that raised the center of gravity excessively.

The development of aircraft engines also led to a considerable improvement in the performance of the G-planes. Gotha already installed 2 engines of 260 hp each and gave the planes an armament of 2 MG (one in front, one in the rear) and soon increased this armament to 3 pieces. The bomb load increased from 150 to 300 kg due to the improved design and towards the end of 1916 to 450 kg,

Above: Gotha G.IV (Englandflieger, England flyer).

Above: Gotha WD1 (1914), an Avro 503 copy.

bringing a further increase in armament to 4 MG. At the end of 1916, the Gothaer Waggonfabrik delivered 50 aircraft, a squadron specially designed for use against England (4 MG, 450 kg bomb load).

In January 1917, especially for the England flights, the type was substantially improved, since requirements were made for the flight duration of more than 6 hours. Towards the end of 1917, the essential demands made on payload and comfort of crew and aircrew, due to the night flights that had to be made to England because of antiaircraft defenses, led to a new design that would again allow daytime flying at altitudes of 6,000 to 7,000 meters. This goal was worked on until the end of the war.

The construction of R-planes, which began in Gotha and was carried out in conjunction with Count Zeppelin, later led to the founding of a special company, Flugzeugwerft GmbH Staaken (see that chapter).

Seaplanes:

Although there was no dedicated water area near Gotha for testing seaplanes, construction was nevertheless initiated in 1913, as already mentioned.

The first seaplane, type WD1, was built, a biplane equipped with a 100 hp Gnom engine (14 cylinders). In February 1914, this aircraft was flown into Warnemünde. Later, by agreement, the city of Rostock built factory facilities and hangars at Breitling[1], which could already be used at the time of the first naval competition for seaplanes (August 1-10, 1914). The WD1 type with a 100 hp Mercedes D.I engine went into production.

For this competition, the company already provided a second type, the WD2 with 150 hp Rapp Rp.III engine.

At the outbreak of the war, which prevented the realization of the competition, the company's branch was taken over by the Naval Administration and formed the core of the Warnemünde Naval Air Station.

The seaplanes intended for the competition were accepted after fulfilling the acceptance conditions, and a larger number, whose delivery took place at the end of 1914, were built. In the spring of 1915, additional seaplane types WD3, WD5, WD7 and UWD were built.

Above: Gotha WD11 (1917), loading a torpedo.

Above: Gotha WD2 for Turkey.

WD3 was a double fuselage biplane, equipped with MG, bomb dropping device and radio equipment (F.T. system).

WD5 was a wartime fast seaplane with 160 hp Daimler engine intended for reconnaissance and bombing.

WD7 and UWD represented the first large aircraft ordered by the Navy. Type WD7 was a biplane with two 120 hp Daimler engines (thrust propeller arrangement) mounted on lower wings; armed with 1 MG. Type UWD, also a biplane with fuselage arrangement between the upper wings. As a result of this design, the two 160 hp Daimler engines could be moved together almost to the point where the airscrew circuits touched.

At the instigation of the Reichs-Marine-Amt, a Type

Gotha

Above: Gotha WD14 (Marine No. 801).

Above: Bomb attachments on a Gotha Go.Va.

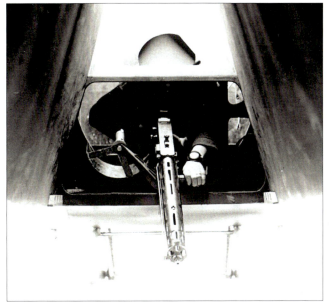

Above: Gotha G.Va, rear MG stand with firing tunnel in the bottom of the fuselage.

Above: The last Gotha developments before the end of the war: Gotha WD 27 on floats, left Bomber Go.X with undercarriage.

WD8 single-engine seaplane with a 260 hp Maybach engine was produced. It was similar to the Type WD7, but was equipped with a machine gun and bombing system.

At the further instigation of the Reichs-Marine-Amt, the WD2 type was modified so that it could also take off and land from land, i.e. it was an amphibious aircraft. The reason for this modification was the fact that seaplanes were to be delivered to Turkey, but Romania did not allow the overflights. It was for this reason that, in order to carry planes by air from Hungary, a wheel axle was added under the floats, and a wheel on either side of each float. The bottom of the float was designed and used as a skid in the last stage. These design features meant that the aircraft could now take off from Gotha airfield for acceptance flights.

The need for fighter seaplanes led to the design of the WD9 type with a 160 hp Daimler engine. Special modification regarding the bracing was necessary so that the MG could shoot forward through the wings past the propeller. The float frame formed the supporting body for the cells. The design of this type of aircraft for bombs and radios resulted in the WD13 type.

The particular success of the G-planes led to the construction of torpedo planes at the instigation of the Navy Department. The types WD11 and WD14 were developed.

WD11 was initially a biplane based on the proven WD8 type with 2 x 160 HP Daimler engines, mounted on the lower wings to the left and right of the fuselage. To accommodate the torpedoes, the fuselage was provided with a recess on its lower side. Based on the experience with the type WD11, the WD14 was equipped with 2 x 200 hp Benz engines (train screw arrangement). Depending on the intended use, the internal equipment was made, namely whether the aircraft was to be used for torpedo shooting, bombing, sea mine laying, or for 11- to 12-hour sea surveillance.

Type WD15 was another biplane with 2 x 160 hp Daimler engine (traction propellers), equipped with bombing equipment and radio.

Production Supervisors (Bauaufsicht): #9

The production supervisors consisted of two officers, five engineers and foremen, 14 service personnel, and two uniformed airmen.

War Material Found by the Inter-Allied Commission of Control (IACC) During Inspections After the End of the War:

25 aircraft
12 Engines
2 compressors for aircraft engines and in addition a large number of spare parts.

Development of Aircraft Factories

Situation after WWI:

In 1919, Gothaer Waggonfabrik resumed its pre-war business of building and repairing railroad coaches. At the end of 1920, it began repairing locomotives.

Whilst Germany was prohibited from military aircraft manufacture by the Treaty of Versailles, Gotha returned to its railway endeavors, but returned to aviation with the rise of the Nazi government and the abandonment of the Treaty's restrictions.

Endnote

1. "The Breitling is a lagoon-like extension of the Lower Warnow River, about 2500 meters wide, just before it flows into the Baltic Sea."

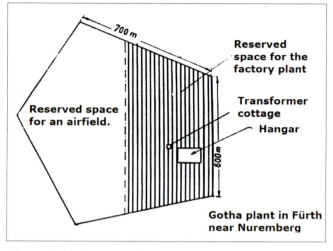

Above: Subsidiary in Fürth (Bavaria) was under construction late in 1918.

Above: Gotha A.I (LE 1 – LE 4), above LE 3 (1913/14)

Above: Gotha B.II (LD2 – LD6), above LD5 (1914)

Above: Gotha B.I (LD7) (1914)

Gotha G.II (1916)

Above: Gotha G.I (GUH) (1915)

Above: Gotha G.III (1917)

Above: Gotha G.IV (1916/17)

Gotha

Above: Gotha G.V (1916/17)

Above: Gotha G.Va (1917)

Above: Gotha G.Vb (1917/18)

Above: Gotha G.VI (1917/18)

Above: Gotha G.VII/GL.VII (1918)

Above: Gotha G.VIII/GL.VIII (1918)

Above: Gotha G.IX (1918)

Above: Gotha G.X (1918)

Above: Gotha WD1 (1913/14)

Above: Gotha WD2 (1915/16)

Development of Aircraft Factories

Above: Gotha WD3 (1915)

Above: Gotha WD4 (UWD), Ursinus-Wasser-DD (1916)

Above: Gotha WD5 (1915/16)

Above: Gotha WD6 (HaBra NW(Go)) (1916)

Above: Gotha WD7 (1915/16)

Above: Gotha WD8 (1916)

Above: Gotha WD9 (1916)

Above: Gotha WD10 (U-1) (1916)

Gotha

Above: Gotha WD11 (1916)

Above: Gotha WD12 (1917)

Above: Gotha WD13 (1917/18)

Above: Gotha WD 14 (1917)

Above: Gotha WD15 (1917)

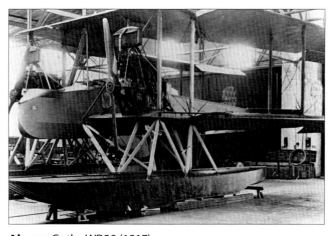

Above: Gotha WD20 (1917)

Above: Gotha WD22 (1918

Above: Gotha WD27 (1918)

4.13 Halberstädter Flugzeugwerke GmbH, Halberstadt (Halb)

Above: Map of Germany and Halberstadt.

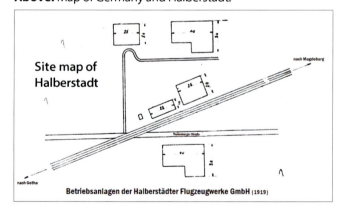

Foundation:

In December 1911, the "Halberstädter Flugplatzgesellschaft" was founded. This paved the way for the construction of an airfield. The city also had the foresight to lay a cable line. This made it possible to build an aircraft factory on the edge of the new airfield site.

The fact that Halberstadt became an "aviation town" was mainly due to the efforts of Max Heckel, a mining councilor.

It can be assumed that Heckel also thought about creating an aircraft factory or at least an aircraft workshop. He was very well acquainted with the owners of the Oschersleben-based company Behrens & Kühne, which was already building novel and transportable aircraft tents and airship hangars in 1911. When Hermann Behrens demonstrated his products in Berlin, he was asked to add light aircraft for the German army headquarters to the company's production program in addition to the production of tents. These were primarily Bristol monoplanes, which were considered the best machines at the time and had been successfully flown by British pilots at Döberitz, the heart of German "military aviation". Behrens therefore contacted Bristolwerke (The British and Colonial Aeroplane Comp., Ltd.), which had only been founded in 1910. He succeeded in concluding an agreement that Bristol aircraft could also be built in Germany.

This company was then founded on April 9, 1912 under the name "Deutsche Bristol-Werke GmbH". After

Halberstadt

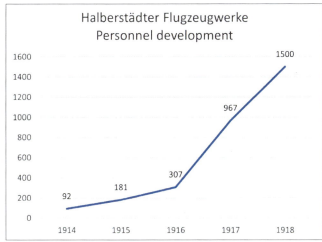

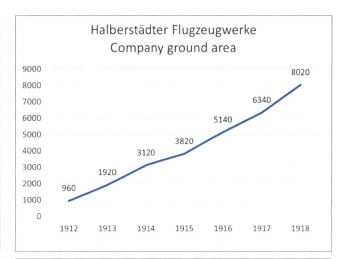

Above: Halberstadt B.I (1914).

the demonstration of the first Bristol aircraft, the Army Administration commissioned four aircraft, but required that the HFW simultaneously train pilots for the German Army. Thus, the Bristolwerke were also among the flying schools financed by funds from the Nationalflugspende.

In September 1914, the company was renamed Halberstädter Flugzeugwerke GmbH. The founders were H. Behrens, G. Behrens, T. Stockhausen and Eduard Schnebel. The "Halberstädter Militär-Fliegerschule GmbH" and Plant II in Klusstraße, where mainly boat building was carried out, were affiliated as subsidiaries.

Factory Facility and Airfield:
From small beginnings, the factory complex on the parade ground, located in front of the Thekenbergs, which also served as an airfield, developed as follows. In terms of working areas were available:
The airfield during the war was located at the Halberstadt racecourse (Harsleberstraße). An increase in aircraft orders also made it necessary to expand the factory facilities, especially from 1916 onwards, so that the company soon had Plants I to IV at its disposal.

Initially, the "Wulfertshöhe" site was added, which housed the fuselage construction department. In the former Lüdde furniture factory, the joinery workshops were expanded.

Staff and Designers:
The company's designer was Dipl.-Ing. Karl Theis, who had previously been employed at Albatroswerke in Johannisthal. At the end of the First World War, around 1,500 workers were employed under his management, with an increasing number of women in the years 1917/18.

Fabrication:
The order situation was very good in view of the German air war armament. With a unit price of 20,000 marks for a (Rumpler) Taube (1912), and 25,000 for the same type in 1914, it was possible to operate profitably. In August 1913,

Aircraft Built by Halberstädter Flugzeugwerke GmbH, Halberstadt (Halb)

Type	Engine [hp], Manufacturer	Crew	Armament	License Built By	Year	Notes
Taube I – IV	100, Merc. D.I; 50, Argus As; 75, Merc. DF80	2	-		1913/14	~20
Halb A.I	100, Merc. D.I	2	-		1914	~5
Halb A.II	80, Oberursel U.0 100, Oberursel U.I	2	-		1914	~20
Halb B.I	80, Oberursel U.0	2	-		1914	50
Halb B.II	100, Merc. D.I	2	-		1915	~10
Halb B.III	120, Merc. D.II	2	-		1915	~10
Halb C.I	100, Oberursel U.I	2	1 MG		1917	1 prototype, armed B-type
Halb CL.II	160, Merc. D.III 190, Argus AS.IIIa	2	1 MG	Bfw: (100) Marta	1917	800
Halb C.III	200, Benz Bz.IV	2	1 MG		1917	6
Halb CL.IV	170, Merc. D.IIIa	2	2 MG	Rol (150)	1918	600
Halb C.V	200, Benz Bz.IVü	2	1-2 MG, camera, FT	Dfw: 200 Bay: 400 Av: 150		
Halb C.VI	150, Benz Bz.III	2	-		1918	prototype
Halb C.VII	260, Maybach Mb.IVa	2	2 MG		1918	prototype
Halb C.VIII	260, Maybach Mb.IVa	2	2 MG		1918	prototype
Halb C.IX	230, Hiero HIV	2	2 MG		1918	prototype
Halb CLS.I	180, Merc. D.IIIaü	2	2 MG		1918	prototype
Halb D.I	100, Merc. D.I	1	1 MG		1916	~3 (25)
Halb D.II	120, Merc. D.II	1	1-2 MG	Han (30), Av (30)	1916/17	~27 (200)
Halb D.III	120, Argus As.II	1	1-2 MG		1916/17	50 (60)
Halb D.IV	150, Benz Bz.III	1	2 MG		1916	prototype
Halb D.V	120, Argus AS.II(O)	1	1-2 MG			57 (120) (12 Turkey)
Halb G.I	2x160, Merc. D.III	3	2 MG		1915/16	1 test a/c
DFW C.V(Halb)	200, Benz Bz.IV	2	2 MG		1916/17	150

86 workers and employees were already employed, and by November 1913, 42 aircraft had been completed, all with propellers from the company's own production.

Initially, the factory produced replicas of the Bristol Box Kite and the Bristol Prier monoplane, but later it also produced its own developments.

The company produced in its own plant almost all the parts for the particularly light flying machines that came into production. By the end of the war, the workforce had grown to about 1,500 workers, resulting in a monthly production of an average of 120 ready-to-fly field-equipped fighter and combat aircraft.

Around the year turn 1916/17, it was characteristic that a top group emerged within the German aircraft factories, which gained a significant lead over other companies, especially with regard to its growth and development activities. In aircraft production, a continuous development of unit numbers took place, not least due to the fact that more and more male and female workers were hired.

Halberstadt

Above: Airfield in Halberstadt with hangar I.

Above: Postcard with different views on the Halberstädter Flugzeugwerke.

Aircraft Development:
The first new development to leave the Halberstadt aircraft works in 1913 was the so-called "Halberstadt-Taube" with a 70 hp engine. The Bristol monoplane (also known as the "Bristol pig"), which had been built up to that point, turned out to be difficult to steer. The Halberstadt Taube, on the other hand, was more cumbersome, but was smoother in the air. In addition, it was more stable (due to the so-called "Röver fuselage").

The first in-house design launched by the company was a biplane with an 80 hp rotary engine under the designation Halb B.I, on which flights were made for the first time in 1914. This design was followed by the Halb B.II type with a stationary engine, on which engineer Voigt, employed as the company's chief pilot, made his first flights, and the first with an aircraft with a stationary engine. In 1916/17, Halberstadt Aircraft Works delivered a series of 85 Halb D.II, Halb D.III and Halb D.V single-seater fighters for the German front lines. In 1917/1918 the company built the first light C-airplane, type Halb CL.II, as a light fighter two-seater. The advantages of this aircraft, besides great maneuverability and speed (168 km/h), were its climbing ability, for it climbed to 5,000 m in 37 minutes. This aircraft went into serial production in 1917 and 80 to 90 were produced per month, and as many as 100 in early 1918. As a further development, the Halb CL.IV type was created. This aircraft was even more maneuverable than the CL.II and reached 5,000 m in 31 minutes during competitive flights at Adlershof. The speed in straight flight was stated at 175 km/h at the time. The next type, the Halb C.V, must be mentioned as another major advance in C aircraft. This type of aircraft was equipped with a 200 hp Benz engine (supercharged), possessed particularly great maneuverability and extraordinarily increased climbing capabilities, which according to the results of the C-machine competition in Adlershof amounted to 5,000 m in 23 minutes. With this performance, the machine was superior even to the C machines with 260 hp supercharged Maybach engines. The horizontal speed was around 180 km/h.

Above: Halberstadt logo.

Production supervisors (Bauaufsicht): 11
Production Supervision No. 11 consisted of an officer, two civilian employees, several crews, and 2 acceptance pilots.

War Material Found by the Inter-Allied Commission of Control (IACC) During Inspections After the End of the War:
195 aircraft
44 engines.

Situation after WWI:
At the end of 1920 the factory employed about 600 workers with the production of agricultural machinery and various sheet metal products (boxes, scales, etc.).

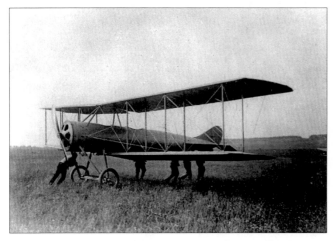

Above: Early Halberstadt biplane (1913), designed by Röver. The "Wickelrumpf" fuselage technology was used.

Above: 4-wheel landing gear on a Halberstadt Taube II.

Above: March 23, 1918 - 1000th HFW aircraft: Halb CL.II.

Above: Halberstadt D.IV Prototype.

Above: Halberstadt D.I, experimentally equipped with rockets.

Above: Halberstadt-Bristol-Taube (1913/14)

Above: Halberstadt A.I (E14) (1914).

Above: Halberstadt A.II (F14) (1914).

Halberstadt

Above: Halberstadt B.I (1914).

Above: Halberstadt B.II (1915).

Above: Halberstadt B.III (1915)

Above: Halberstadt C.I (1917).

Above: Halberstadt CL.II (1917)

Above: Halberstadt C.III (1917)

Above: Halberstadt CL.IV (1918)

Above Right: Halberstadt C.V (1918)

Right: Halberstadt C.VI (1918)

Development of Aircraft Factories

Above Right: Halberstadt C.VII (1918)

Above: Halberstadt C.VIII. The C.VII and C.VIII were identical except for wing span. The C.VIII had longer span, which gave it better climb and flying qualities at altitude. (1918)

Above Right: Halberstadt C.IX (1918)

Above Right: Halberstadt CLS.I (1918)

Above: Halberstadt D.I (1916)

Above: Halberstadt D.II (1916)

Above: Halberstadt D.III (1916)

Above: Halberstadt D.IV (1916)

Halberstadt

Above: A Schlasta of Halberstadt CL.II attack aircraft.
Left: Halberstadt D.V (1916/1917)

Above: DFW C.V(Halb) (1916/1917)

Above: Halberstadt G.I (1915/1916)

Above: Halberstadt CL.IV (1918)

4.14 Hannoversche Waggonfabrik A.G., Hanover (Han)

Above: Advertisement 1913.

Foundation:
The Hannoversche Waggonfabrik was founded in 1871 by Messrs. Max Menzel, Buschmann and Holland and became a joint stock company in 1898 with its headquarters in Hanover-Linden.
Aircraft construction began in the fall of 1915. Hannoversche Waggonfabrik A.G. was given the abbreviation "Han" by the Flugzeugmeisterei.

Factory Facility and Airfield:
The company erected various buildings at its own large airfield for the purpose of aircraft production:

Autumn	1915	2,100 m²
April	1916	3,600 m²
October	1916	5,600 m²
April	1917	7,000 m²
October	1917	12,500 m²
December	1917	14,000 m²

Fabrication:
The company, founded only after the beginning of the World War, was initially engaged in the production of C aircraft

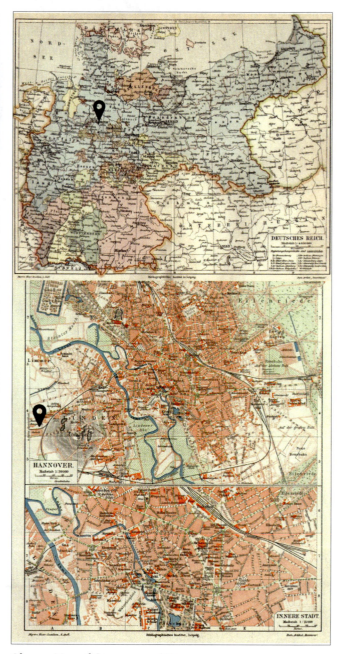

Above: Map of Germany and Hannover.

under license (Aviatik and Halberstadt), and later also D aircraft by Rumpler.

In the middle of 1917, Hannoversche Waggonfabrik started the production of a self-designed aircraft, type Han CL.II, and executed the same with different engines. The type known as CL.IIa was built under license by LFG (Han CL.IIa(Rol), see there). In addition, the company was engaged in the development of its own C-aircraft with a 260 hp Maybach engine. In addition, there was work on a fighter two-seater.

Hannover

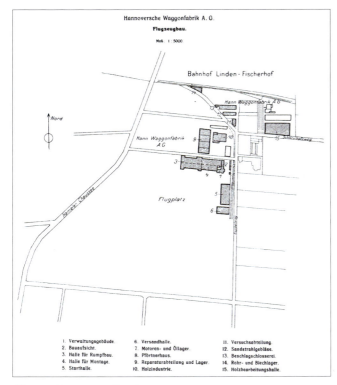

Above: Hannover site plan.

Above: HaWa hangar, final assembly hall.

Above: Loading and transport of finished HaWa aircraft to the front.

Monthly production developed as follows:

April 1916	3
October 1916	17
April 1917	18
October 1917	60
January 1918	100

After the armistice, it delivered 200 unfinished aircraft to the "Unterstell-Kommando" (Storing Department), including:

- 100 Han CL.III
- 38 Han CL.IIIa
- 57 Han CL.V
- 5 Han CL.VM

Staff and Designers:

The company's employees have developed as follows:

	Workers	Officials
April 1916	300	10
October 1916	425	22
April 1917	1.200	52
October 1917	1.700	115
January 1918	1.800	140

Initially, the engineer Mairich was employed as the designer of Hannoversche Waggonfabrik A.G., followed by Dipl.-Ing. Hermann Dorner from September 1916. The commercial director of the plant was Rudolf Stahlschmidt, and the technical director was Regierungsbaumeister Albrecht Nuß.

Above: Loading and transport of finished HaWa aircraft to the front.

Above: Loading and transport of finished HaWa aircraft to the front.

Development of Aircraft Factories

Aircraft Built by Hannoversche Waggonfabrik A.G., Hanover (Han)

Type	Engine [hp], Manufacturer	Crew	Armament	License Built By	Year	Notes
Han D.I originally: Halb D.II(Han)	120, Merc. D.II	1	1 MG		1916/17	30
Han CL.II	180, Argus As.III(O)	2	2 MG	LFG (Rol): ~50 CL.IIa	1917	446
Han CL.III	160, Merc. D.III	2	2 MG		1917	80 + 100 postwar
Han CL.IIIa	180, Argus As.III(O)	2	2 MG		1917/18	707 + 38 postwar
Han CL.IIIc	190, NAG C.III	2	?		1917	?
Han C.IV	245, Mayb. Mb.IVa	2	2 MG		1918	2 prototypes
Av C.I(Han)	150, Merc. D.III	2	1 MG		1915	146
Ru C.Ia(Han)	180, Argus As.III	2	2 MG		1916/17	375
Han CL.V	170, Merc. D.IIIa / 185, BMW BMW.IIIa	2	2-3 MG		1918	46 + 62 postwar (incl. 5 CL.VM)

Aircraft Development:
In the summer of 1915, due to its large wood reserves, the company initially began building propellers, to which it soon added a repair station of aircraft surfaces for Fea 5, Hanover. It was only in November 1915 that the company began building new aircraft, under license from Automobil- und Aviatik-A.G., Leipzig-Heiterblick. This first order was followed by another order under license Halberstädter Flugzeugwerke and this in turn by a next license order for Rumpler aircraft. In addition to the production of these licensed aircraft, the company was engaged in an airplane of its own design and brought it out in July 1917. This aircraft stood out with its good flight performance, and was subsequently delivered in large numbers.

Production Supervisors (Bauaufsicht): #10
Production Supervision No. 10 was installed as of September 18, 1916. It consisted of 17 men, including two engineers and two post-flight pilots.

At the beginning of October 1917, an additional armament supervision was established at the company, which consisted of the leader, a clerk and a machine gun technician.

War Material Found by the Inter-Allied Commission of Control (IACC) During Inspections After the End of the War:
3 civilian aircraft
38 fuselages of Han CL-Types

Situation after WWI:
In 1919, Hannoversche Waggonfabrik (HaWa) resumed its pre-war production: Cargo wagons and passenger coaches. Extensive expansion work was carried out for this purpose in August 1920.

Above: Han CL.IIIa (1917/18).

Above: Representatives of HaWa and members of the military authorities celebrate the delivery of the 1000th Han CL.II aircraft.

Above: Hannover Han C.III after a taxi accident with an Alb D.V.

Hannover

Above: HaWa airfield was used postwar for civilian purposes.

Above: Hawa F6, disarmed CL.V introduced in 1919 as a passenger aircraft, still believing in a continuation of the German aircraft industry.

Above: Hannover CL.II (1917)

Above: Hannover CL.III (1917)

Above: Advertisement 1913.

Above: Hannover CL.IIIa (1917/18)

Above: Hannover CL.IIIc (1917)

Development of Aircraft Factories

Above: Hannover C.IV (1918)

Above: Hannover CL.V (1918)

Above: Halb D.II(Han), also known as Han D.I (1916/17)

Above: Aviatik C.I(Han) (1915)

Above: Rumpler C.Ia(Han) (1916/17)

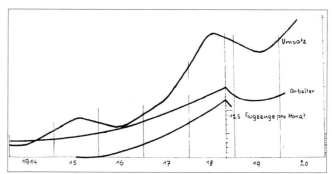

Above: Graphical representation of the development of Hannover sales, personnel, and aircraft production during the war years.

4.15 Hansa und Brandenburgische Flugzeugwerke AG (Brand)

Oberleutnant d. R. Christiansen schießt am 3. Oktober 1917 an der Themsemündung mit einem Flugzeug der **Hansa- und Brandenburgischen Flugzeugwerke A.-G.** ein englisches doppelmotoriges Curtiß-Flugboot ab.

Above: Advertisement 1917.

Above: Map of Germany and Briest.

Above: Brandenburg KDW in front of the H.B.F. hangar in Hamburg-Fuhlsbüttel.

Foundation:

The Hansa und Brandenburgische Flugzeugwerke GmbH (later Corporation (AG)), was a major supplier of aircraft to both the German and Austro-Hungarian Military and Naval Forces during the War of 1914-1918. Four years after it was established in the Winter of 1914, the Company had become the chief unit in a complex of aircraft and engine plants which encompassed all the countries of the Central Powers.

The well-known Austrian aviation pioneer Igo Etrich, who had made a name for himself by building the Etrich Taube, founded the Brandenburgische Flugzeugwerke GmbH in Briest near Brandenburg an der Havel in 1914 with Ernst Heinkel, who had already worked for him before, as chief designer, initially to exploit the patents of the Lohner Boot (Austrian seaplane with fuselage).

In early 1914 Etrich went into partnership with Gottfried Krüger, a small industrialist of Brandenburg. As a result of this amalgamation the "Brandenburgische Flugzeugwerke GmbH" was born, with the main factory at Briest (Brandenburg) and an assembly plant (the original Etrich factory) at Oberaltstadt (heute Horní Staré Město, Tschechien).

In February the Hansa-Flugzeugwerke Carl Caspar A. G. of Hamburg-Fuhlsbüttel was taken into the partnership. The title of the firm was changed once more to "Hansa-und-

Development of Aircraft Factories

Aircraft Built by Hansa und Brandenburgische Flugzeugwerke AG (Brand)

Type	Engine [hp], Manufacturer	Crew	Armament	License Built By	Year	Notes
Landplanes						
Brand D	110, Benz Bz.II	2	–		1914	~12
Brand FD	100, Argus AS.I	2	–		1914	min. 3
Brand LDD* (Austrian B.I)	100, Merc. D.I	2	–		1915	
Brand DD* (Austrian C.I)	160, Merc. D.III 160, AD Dm160	2	2–3 MG	Phönix: 400 UFAG: 834	1915/17	80, built at Rin, completed at Brand
Brand MLD	150, Benz Bz.III	2	1 MG		1915	prototype
Brand ZM	2x160, Maybach Mb.III	3			1915	2 prototypes
Brand GF* (Austrian G.I)	2x160, AD Dm160	3	2 MG, Bombs	UFAG: 12	1915	27
Brand KD* (Austrian D.I)	160, AD Dm160 150, AD Dm150 150, Benz Bz.III 185, AD Dm185	1	1 MG	Phönix: 71	1916	50
Brand KF	150, Benz Bz.III	2	1 MG		1916	prototype
Brand KDD	160, Merc. D.III 160, AD Dm160	2			1916	2 prototypes
Brand K* (Austrian C.II)	185, AD Dm185 200, AD Dm200 185, Merc. D.IIIa	2	1 MG		1916	2 prototypes
Brand L14	200, Hiero Typ H	1	2 MG		1917	prototype
Brand L15	350, AD Dm345 300, AD Dm300	2	2 MG		1917	prototype
Brand L16	185, AD Dm185	1			1917	prototype
Alb B.I	160, Merc. D.III	2				On behalf of Phönix for Austria

* also manufactured in Austria-Hungary by Phönix Flugzeug-Werke AG, Fliegerarsenal Flugzeugwerke and/or Ungarische Flugzeugfabrik AG under license.

Brandenburgische Flugzeugwerke GmbH". Ernst Heinkel was appointed Chief Designer in March and it seemed that the Company was at last on a stable foundation.

In July 1915, the ambitious Austrian industrialist Camillo Castiglioni acquired Brandenburgische Flugzeugwerke GmbH and Hansa-Flugzeug-Werke Hamburg Carl Caspar in Hamburg-Fuhlsbüttel, which he merged with his company Deutsche Aero-Gesellschaft AG, which had been in existence since 1914, and renamed all together in Hansa-und Brandenburgische Flugzeugwerke AG. In 1918, Max Oertz's boat and yacht shipyard in Hamburg was added, apparently to further expand the production of seaplanes.

Owing to the ever-increasing demand for more aircraft, the Government decreed that all technicians with aeronautical experience should be drafted back into the industry. Under this edict, Karl Caspar was released from the Army in late 1916 and in accordance with his agreement with Brandenburg he decided to withdraw from the Castiglioni Group.

The Fuhlsbüttel plant then became the "Hanseatischen Flugzeugwerke Karl Caspar AG" and commenced licence production of the Albatros C.III. Several hundred of these machines were produced in the next few months and later, in 1918, the firm manufactured the Friedrichshafen G.IIIa. About 93 machines had been built when the Armistice terminated activities.

The subsidiary plant in Berlin-Rummelsburg, founded in 1916, became an independent company in 1917 under the

Brandenburg

Aircraft Built by Hansa und Brandenburgische Flugzeugwerke AG (Brand)

Type	Engine [hp], Manufacturer	Crew	Armament	License Built By	Year	Notes
Floatplanes						
HaBra AE (Lohner T)	?	2	–		1915	1
HaBra FB	150, Benz Bz.III	3	1 MG		1915/16	6+11 for Austria
HaBra CC	150, Benz Bz.III, 185, AD Dm185 or 200, Hiero Type H	1	2 MG	Phönix (35)	1916/17	36 + 37 for Austria
HaBra W	150, Benz Bz.III	2	–		1914/15	25
HaBra NW	160, Merc. D.III	2	50 kg Bombs	Gotha: 30	1915	HaBra: 34 +5 for Austria
HaBra GNW	160, Merc. D.III	2	10 Bombs of 5 kg		1915/16	16
HaBra LW	160, Merc. D.III	2	1 MG		1916	1 (3)
HaBra GW (based on HaBra G.I)	2x160, Merc. D.III	3	1–2 MG 1 Torpedo		1916/17	21 (26)
HaBra GDW	2x200, Benz Bz.IV	2	1 MG 1 Torpedo		1917	1 prototype
HaBra KW	200, Benz Bz.IV	2	1 MG		1916/17	3
HaBra KDW	150, Benz Bz.III 160, Mayb. Mb.III	1	1–2 MG		1916/17	58 + 10
W.11	200, Benz Bz.IV	1	2 MG		1917	3
W.12	160, Merc. D.III 150, Benz Bz.III 190, Benz Bz.IIIb	2	2 MG		1917	147
W.16	160, Oberursel U.III	1	2 MG		1917	2–3
W.19	260, Mayb. Mb.IVa	2	2 MG		1917	>77, some with cannon
W.20	80, Oberursel U.0	1			1917	3 (Germany)
W.25	150, Benz Bz.III	1	2 MG		1917	1
W.26	260, Merc. D.IVa	2	2 MG		1918	3
W.27	195, Benz Bz.IIIb	2	2–3 MG		1918	1–3
W.29	150, Benz Bz.III	2	2–3 MG		1918	78
W.32	160, Merc. D.IIIa	2	2–3 MG		1918	3 (of 5)
W.33	300, BuS IVa 260, Mayb. MB.IVa	2	2 MG		1918	7 (+ 19 postwar) 28 Norway 122 Finland
W.34	300, BuS IVa	2	2 MG		1918	1 (of 3)

trade name Flugzeugwerke Albert Rinne (see also there).

Fabrication:
During the war, Hansa und Brandenburgische Flugzeugwerke and Flugzeugbau Friedrichshafen, were the two largest seaplane manufacturers in Germany, with Hansa und Brandenburgische Flugzeugwerke specializing in particular in the construction of reconnaissance and combat seaplanes. By contrast, the Prussian army aviators' use of land-based aircraft was small compared to the share of other German aircraft manufacturers.

In all, they built 600 aircraft (in other publications a number of 1000) for the German and associated Navys. During the war, the Flugzeugmeisterei gave the company the short name "Brand".

Aircraft Built by Hansa und Brandenburgische Flugzeugwerke AG (Brand)

Type	Engine [hp], Manufacturer	Crew	Armament	License Built By	Year	Notes
Flying Boats						
FB	165, AD Dm165 175, Rp 175, 200, Hiero Type H	2–3	1 MG	UFAG Type K: >60 (Austria)	1916	6
W.17	200, Hiero Type H	1			1918	1-2 prototypes
W.18	150, Benz Bz.III 200, Hiero Type H 230, Hiero Type H_{IV}	1	2 MG		1917/18	47 (+9) for Austria
W.20	80, Oberursel U.0	1	–		1917	3
W.23	160, Merc. D.III	2	2 MG		1917/18	3
W.35	300, Basse&Selve 260, Mayb. Mb.IVa	4	1 cannon Bombs		1918	0

Staff and Designers:
The personal was increasing constantly as shown hereafter. **Ernst Heinkel** was offered a position with Luft-Verkehrs-Gesellschaft („LVG") at Johannisthal, Berlin. He accepted the job in October 1911, and several weeks later was joined by the Swiss engineer **Franz Schneider**, who became Chief Designer. Schneider's experience had been gained at the French firm of Nieuport and his designs bore a marked resemblance to those of his previous employer; but whereas the French planes were adequate for their purpose, Schneider's aircraft were considered somewhat unsafe to fly. Sometime in 1915 Deutsche-Aero Gesellschaft became an Associate Company of the Brandenburg Consortium and the firm's Rummelsburg factory, near Berlin, received a share of Hansa-Brandenburg's contracts. **Dr. Walter Lissauer**; the Chief Engineer of the Company, was appointed Technical Adviser to the Castiglioni Group, Lissauer was one of the leading figures of the early German aeronautical industry. Trained as a Physicist he had learned to fly in 1910 (Certificate No. 22). After working briefly for Kühlstein Wagenbau AG in Berlin, he had joined the Fokker Aeroplanbau at Schwerin where he organized the Fokker Military Flying School.

Aircraft Development:
According to the company's own information, about 600 aircraft were delivered at Briest near Brandenburg. The Imperial Navy was the main supplier.
For the army, the Brand C.Ib was used in the war.

Heinkel 's first design for Brandenburg was completed in April. A two-seater reconnaissance biplane, it was powered with a Benz Bz.II engine of 110 hp. Designated type "D", some 12 aircraft were eventually produced, four of which were supplied to Austro-Hungary in late 1914.

In mid-1914 an enlarged version of the Type 'D' was fitted with pontoons and designated Type "W". The German Navy, after testing the prototype, ordered 24 machines powered with the Benz Bz.III of 150 hp. Subsequently, 2 further aircraft were built which were fitted with the Maybach Mb.III of 160 hp. and in early 1915, one experimental machine powered with the Argus As.II of 140 hp.

In the Autumn of 1914, the new Brandenburg Type "FD" was ready for tests. A development of the Type "D", the prototype was powered with the 100 hp Argus As.II. After testing, at least 8 other machines were constructed with similar power plants. These aircraft had no Military Number and were probably the property of Hansa-Brandenburg, being used by the Caspar Flying School at Fuhlsbüttel.

In early 1915 the Type "NW", a coastal patrol seaplane, was produced at the Briest works. 32 machines were built there and a repeat order for 30 additional aircraft were sub-contracted by Gothaer Waggonfabrik AG in early 1916.

In the Spring of 1915, the Type "LDD" was ordered in quantity by the Austro-Hungarian Air Service. Machines were required urgently, but production capacity at the Briest plant was fully utilized. An Albatros B.I production line at Phönix was therefore changed over to the Brandenburg type after about the 25th machine and the initial order filled by the Austrian plant.

Second marine aircraft type for 1915 was the Brandenburg "LW", a float-equipped biplane powered with a Mercedes D.III of 160 hp. This was an experimental type which was fitted with a machine gun - one of the first installations of this weapon on a seaplane. The sole machine (Navy No; 571) was tested late in the Summer but no production was undertaken.

The Type "DD" was a cleaned-up version of the earlier "LDD". Fitted with a Mercedes D.III of 160 hp, the performance was so improved over the latter type that on September 21st, 1915, it set a world altitude record, with 4 passengers, of 4760 m. Three machine guns were fitted, one

Above: Production line in Brandenburg (Briest).

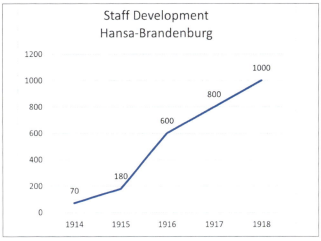

pivot-mounted and two fixed to fire over the wing.

In the Autumn of 1915 Heinkel's first flying boat made its debut. The aircraft was designated type "FB" and was powered by a Benz Bz.III of 150 hp.

Brandenburg had much greater success with the floatplane version of the "GF". This was designated Type "GW" and was the first torpedo-bomber in the German Navy. 21 machines were ordered in 1916 in both long and short nosed versions all being powered with two Mercedes D.III's of 160 hp each.

The C.I (Brand DD) was, without doubt, the most common Brandenburg type in the Austro-Hungarian Air Service. Between 1916 and 1918 about 834 were produced by UFAG. The Briest plant delivered another 85 machines, and Phönix built about 400.

A more powerful version of the Type "GW" appeared in mid-1916. This aircraft, the Type "GDW" had a greater wingspan than its predecessor and more powerful engines (2 Benz Bz.IV's of 200 hp. each). It was also tested with the Benz Bz.IVa's of 220 hp but as no orders were placed by the Navy it remained the sole example.

After the parent factory had started deliveries of the "GW" Series in 1916, Heinkel's team produced a floatplane with the designation "KW". This was a development of the "LW" and was powered with a Benz Bz.IV of 200 hp (some sources quote the Benz Bz.IVa of 220 hp). Three aircraft were supplied as training machines, numbered 588-590.

Just after mid 1916 Heinkel's team of marine aircraft designers had their greatest success thus far when the Brandenburg Type "KDW" was ordered in quantity for the German Navy. 58 examples were constructed, utilizing the Benz Bz.III engine of 150 hp and 10 aircraft were delivered with the more powerful Maybach Mb.III of 160 hp. One further Maybach powered example with increased wingspan may also have been built.

Heinkel's next aircraft was a single seat flying boat and as a tribute to Castiglioni it was given the designation "CC". 36 of these aircraft were supplied to the Navy, all with Benz Bz.III engines of 150 hp.

Following the "CC" all subsequent Brandenburg aircraft for the Navy were designated by the prefix letter "W" followed by a number. Thus, the next design was Type W.11, and this was a larger and more powerful version of the "KDW". Three were ordered by the Navy, but as performance proved to be inferior the last machine may not have been built.

Although the Type W.12 was completed in December 1916, weather conditions prevented it being tested until the following month. This was the first Brandenburg type with a field of fire from the observers' guns which gave effective protection against a rear attack. The German Navy was very satisfied with its performance, and it was ordered into production.

The W.14, W.15, and W.17 were projects which never got beyond the layout phase. Type W.16 was a design for a small "Station Defense" aircraft powered by an Oberursel U.III rotary engine of 160 hp. Three machines were ordered, but it is believed that only two were built in late 1916.

UFAG were unable to build any D.I's as they were fully occupied with the production of the Brandenburg C.I. The 48 aircraft of UFAG's order were therefore supplied by the Briest works. This was the last Brandenburg type supplied in quantity by the German firm, as hence forth the design teams headed by Kirsten & Gabriel of Phönix and Orovecz and Blodek of UFAG were unable to develop their own designs independently of the Briest Design Office.

The second experimental type of 1916 was the Brandenburg "KF" which was also Heinkel's second pusher scout. Powered by a Benz Bz.III of 150 hp, the twin-truss tail booms of the "MLD" were replaced with enclosed booms approximating in size to an actual fuselage structure. No production was undertaken as the performance did not match the complexity of the design.

The third 1916 type was the "KDD" with an Austro-Daimler engine of 160 hp. This was the two-seat reconnaissance version of the "KD" and like that type had the "Star-strut" arrangement for bracing the main surfaces. Two machines were built and sent to UFAG as samples, but no production was undertaken, as it was considered that the strut arrangement was too complex.

Above: Brandenburg W.29. View at radiator, MG and engine valves.

The last 1916 type was the "K" a specimen of which was sent to UFAG as a sample for a proposed replacement series for the "LDD" (C.I.) still being built there. However, the improvement in performance was not sufficient to warrant production of the type and the project was abandoned.

In early 1917 Heinkel designed a small single-seat flying-boat powered with a 200 hp Hiero engine. This aircraft was designated W.18 and the German Navy ordered 3 for trial purposes. However, only the prototype is believed to have been constructed, as the Navy was never very keen on flying-boats, preferring the more easily maintained floatplane.

Late in the year a redesigned version of the Type W.18 was built for the Austro-Hungarian Navy. This machine was powered with the 350 hp Austro-Daimler engine Dm345 and was generally referred to as the Type "KG". 60 aircraft were eventually delivered from Brandenburg, 60 from UFAG and 8 from Phönix. In early 1918 OEFFAG constructed three modified machines, and these were designated Class "T".

In mid-1917 the German Navy decided to equip a submarine with a small scouting plane, and for this purpose Brandenburg designed the Type W.20, which was powered with the 80 hp Oberursel rotary motor. Three examples were built, the first without interplane struts. Another Series was proposed with the 110 hp Le Rhone, but these were never built, as the class of submarine for which both series were intended never went into service.

In the Summer of 1917 Hansa-Brandenburg absorbed the firm of Oertz-Werke GmbH of Hamburg/Reiherstieg. Max Oertz was a Naval architect and engineer who had turned his talents to flying-boat design in 1913. His aircraft were all pusher types with the engines buried in the hull and driving the airscrews through gears and shafting.

Type W.25 was similar to the earlier KDW, except that it reverted to normal interplane bracing instead of the "star-strut" design of that type. It originally had ailerons on the upper wing only, but at a later date both wings were so fitted. Power plant was a Benz Bz. III of 150 hp. It was re-engined

Right: Brandenburg W.20, developed for use from submarines

Above: Brandenburg KD (Austrian D.I). 50 aircraft were built at Brandenburg, 71 at Phönix.

Above: Brandenburg W.19. Test installation of 2cm cannon.

Above: Brandenburg GW, navy number 700 with a torpedo in the torpedo bay.

Above: Front view of Brandenburg CC (1916/17).

Above: This picture shows an experimental flying boat from HBF. The crosswise wing struts indicate a predecessor of the CC.

Above: Hansa-Brandenburg W.29. 78 aircraft were built in 1918.

Above: The Brandenburg FB was Brandenburg's first flying boat, developed in 1916.

later with a Maybach Mb. III of 160 hp.

At the start of 1918 it was apparent that aircraft supplied to the German Navy in the future would be required to carry more payload at greater speed over longer distances; hence installed power would increase, either through increased output of engines or the use of multiple engine installations. It was also clear that more machines would be needed to carry out the increasing tasks which befell a Navy now turning more and more to a defensive strategy.

Orders for aircraft had increased from small batches in 1914 to dozens in 1918, in spite of difficulties in the procurement of materials and components. This delayed deliveries to such an extent that many aircraft were short of vital parts as early as the spring of 1918. However, Hansa-Brandenburg managed to supply and deliver aircraft in fulfilment of their Navy contracts and at the same time produce experimental machines in the hope of receiving orders.

Early in 1918 the Briest plant produced the W.26 which was a long-range patrol seaplane with a 260 hp. Mercedes D.IVa engine. Three were built, but no production followed. The W.27 was a development of the W.12 with a Benz Bz.IIIb of 190 hp. One was built, although two machines were ordered. The W.28 remained in the project stage.

In March 1918, Heinkel's successor to the W.12 was tested at the Brandenburg factory. The W.29 was a low-wing monoplane on pontoons powered by a Benz Bz. III of 150 hp and using many components of the former type for ease of production. The design was an immediate success and over 100 were ordered, although only 78 were delivered before the Armistice terminated production.

Type No. "W.30" was allocated to a large flying boat design which showed the trend in development of German naval aircraft. This machine was to be powered by three Mercedes D.III engines each of 160 hp - the first time such an installation had been proposed. Serial Nos. 2301 and 2302 were reserved, but the project was cancelled in favour of a larger aircraft.

The Type W.32 appeared in early 1918 and was essentially a re-engined W.27. Five machines were ordered, but the latter two were cancelled in April so that the full production resources of the Brandenburg firm could be concentrated on the manufacture of the superior W.33.

The last Brandenburg machine which was produced in any quantity was the W.33 of mid-1918, a larger and more powerful development of the W.29. Twenty-six aircraft had been delivered by November, 23 powered by the Maybach Mb. IVb of 260 hp. and 3 fitted with the Maybach Mb.IVa of 245 hp. (Some sources state that these machines were, in fact, W.29's; that the latter three machines were fitted initially with the 300 hp Basse und Selve BuS.V, and that these engines were later changed to 260 hp Maybachs in order to standardise the power plant. No information on these statements is available).

The W.34 was a slightly larger version of the W.33 and was powered with the 300 hp Basse und Selve BuS.V engine. Six were ordered, but only one example was completed, five were awaiting installation of power plants at the Armistice. The W.34 was the last Brandenburg machine completed in Germany in 1918.

In the summer of 1918 Brandenburg submitted a design for a large flying-boat designated W.35 which was to be powered with two Basse und Selve engines of 300 hp each. These aircraft were to carry a crew of four and in addition to the normal cargo of bombs a Becker-Kanone was installed

Development of Aircraft Factories

Above: The Brandenburg KDW was the last model where only letters were used as the model designation. The following seaplane designations started with a "W".

Above: „By order of the Entente. II. S.F.A. Wilhelmshaven." Destroyed W.29.

Above: "The Rest. II. S.F.A. Wilhelmshaven." Wings, fuselage, engine parts of W.29.

Above: Brandenburg D (1914)

on a swivel mount in the front cockpit. It was hoped that these machines would be the answer to the large Curtiss and Felixstowe boats which were giving some trouble to their smaller and more lightly armed German opponents.

Construction was commenced on the premises of the former Oertz Works, and two hulls were almost completed at the end of October 1918. However, after the Armistice no further work was done on these machines and the unfinished hulls were broken up.

Brandenburg Types W.36 and W.37 were projects and although the form er never got beyond the
layout stage, the latter was eventually built in 1921 as the Caspar S.1 (or He.1).

Production supervisors (Bauaufsicht):
Since Hansa und Brandenburgische Flugzeugwerke GmbH almost exclusively supplied the Imperial Navy with aircraft, production supervision was carried out by the responsible naval office.

War Material Found by the Inter-Allied Commission of Control (IACC) During Inspections After the End of the War:
After the Armistice Brandenburg's Contracts with the Navy were cancelled.

Work had virtually come to a standstill and several dozen airframes in varying stages of completion plus a few new power plants, awaited the attentions of the skilled workmen who would never return.

Early in 1919 the five engineless W.34's on hand at Brandenburg were sold to the recently formed Finnish Air Force and smuggled piece by piece out of Germany. Evidence shows that various aircraft were sold to many other countries, or licenses to reproduce them were granted to various companies abroad.

Late in 1920 the Inter-Allied Control Commission placed a total ban on all German aircraft production and apart from certain items retained for testing, evaluation or historical interest all material and equipment was ordered to be destroyed. (A large quantity of this material had in the meantime "disappeared" into safe hiding places all over Germany).

The IACC report states that 119 seaplanes and 30 engines, coming probably from the factory, were counted in a brick factory located at 3 kilometers from the place where the factory was.

Situation after WWI:
The factory was completely rusticated after the war; only the management headquarters remained. The machines, mostly woodworking, were sold.

Endnote
1 Most of the information of this chapter is taken from D.T. Pardee's article "Hansa-Brandenburg, 1914-1918, A Short History."

Brandenburg

Above: Brandenburg FD (German B.I) (1914)

Above: Brandenburg DD (Austrian C.I) (1915/16)

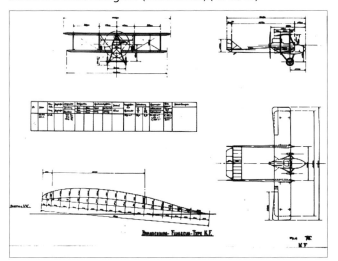

Above: Brandenburg KF (1916)

Above: Brandenburg LDD (Austro-Hungarian B.I) (1915)

Above: Brandenburg MLD (1915)

Above: Brandenburg KDD (1916)

Above: Brandenburg K (Austrian C.II) (1916)

Left: Brandenburg L14 (1917)

Development of Aircraft Factories

Above: Brandenburg L15 (1917)

Above: Brandenburg L16 (1917)

Above: Brandenburg ZM (1915)

Above: Brandenburg GF (Austrian G.I)

Above: Brandenburg FB (1915/16)

Above: Brandenburg CC (1916/17)

Above: Brandenburg W (1914/15)

Above: Brandenburg NW (1915)

Brandenburg

Above: Brandenburg GNW (1915/16)

Above: Brandenburg LW (1916)

Above: Brandenburg GW (1916/17)

Above: Brandenburg GDW (1917)

Above: Brandenburg KW (1916/17)

Above: Brandenburg KDW (1916/17)

Above: Brandenburg W.11 (1917)

Above: Brandenburg W.12 (1917)

Above: Brandenburg W.13 (Austro-Hungarian K) (1917)

Above: Brandenburg W.16 (1917)

Above: Brandenburg W.18 (1917)

Above: Brandenburg W.19 (1917)

Left: Brandenburg W.20 (1917)

Above: Brandenburg W.25 (1917)

Above: Brandenburg W.26 (1918)

Above: Brandenburg W.27 (1918)

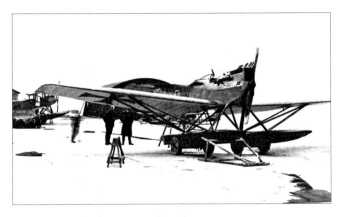

Above: Brandenburg W.29. (1918)　　　　　　　　　**Above:** Brandenburg W.32. (1918)

Above: Brandenburg W.33. (1918)

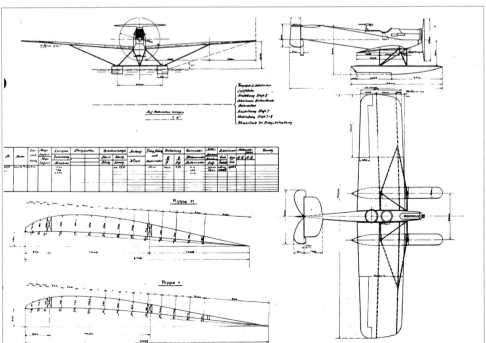

Right: Brandenburg W.34 drawing. (1918)

4.16 Hanseatische Flugzeugwerke AG (Karl Caspar), Hamburg-Fuhlsbüttel (Hansa)

Above: Advertisement 1913.

Foundation:

Karl Caspar[1] learned to fly on Etrich-Taube and received the pilot certificate No. 77. In 1911, at the age of 28, he laid the foundation stone for his company. In the fall of 1911, Mr. Karl Caspar founded the local company, which at that time consisted of a small flight school under the name "Centrale für Aviatik Hamburg" on the Wandsbek parade ground. The transformation into a stock corporation took place on January 1, 1917.

A fire accident in the summer of 1912 led to the company being relocated to a more suitable airfield - Hamburg-Fuhlsbüttel. The Hanseatische Flugzeugwerke were rebuilt there on the basis of a contract with the Hamburg Luftschiffhafen-Gesellschaft, which had already leased another hangar to the Navy.

Caspar was by all means a successful pilot: on June 19, 1912, he set a German altitude record of 3,245 m with a Rumpler pigeon on the occasion of the Nordmarken flight. He achieved further successes at the "Krupp Flight Week 1912" and the "East Prussia Flight 1913".

In 1914, the company name changed from "Centrale für Aviatik Hamburg" to "Hansa-Flugzeugwerke Hamburg Karl Caspar," while the flying school was henceforth known as the "Hanseatische Flugschule". At the outbreak of the First World War, Caspar immediately fulfilled his so-called "army duty" and reported for military service with the Fliegertruppe. In order not to leave his company without management for a long time during wartime, he negotiated a merger of his company with the "Brandenburgische Flugzeugwerke GmbH" in Briest near Brandenburg (Havel) and the "Deutsche Aero-Gesellschaft AG" in Berlin in 1915. The new company was named "Hansa- und Brandenburgische Flugzeugwerke AG," and Berlin was set as

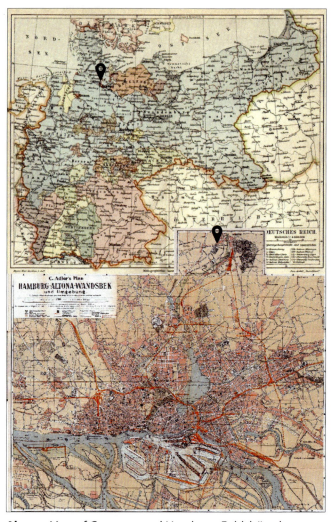

Above: Map of Germany and Hamburg-Fuhlsbüttel

its headquarters, with operating facilities in Berlin, Hamburg and Briest. The main shareholder of the merged plants was the Vienna-based Kommerzialrat Camillo Castiglioni, who also held the position of general manager.

Ernst Heinkel became chief designer and technical director. The former owners Kommerzienrat Gottfried Krüger (Brandenburgische Flugzeugwerke) and Referendar Karl Caspar (Hansa-Flugzeugwerke) were each given a seat on the supervisory board of the new company. The financier of this foundation was Kommerzienrat Castiglioni from Vienna, who also arranged orders and students from Austria.

In the summer of 1916, about half of the factory burned down, but was immediately rebuilt on the same site by the Free and Hanseatic City of Lübeck in an expanded massive style.

During the war, the Hanseatische Flugzeugwerke was given the short name "Hansa" by the Flugzeugmeisterei.

When the Hansa-Flugzeugwerke were incorporated into the new company, Caspar had reserved the right to withdraw his

Caspar

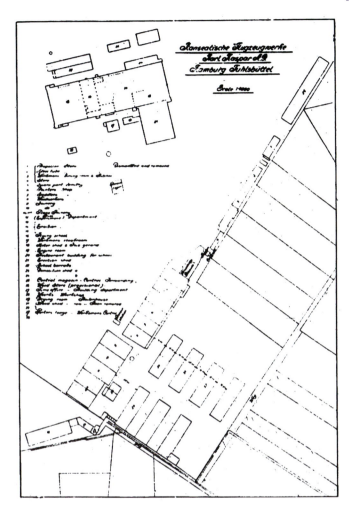

Above: Site plan of Hamburg-Fuhlsbüttel.

Above: Karl Caspar after his high-altitude flight in 1912 (3,245 m) with Gotha-Taube. Engine: Rumpler V8 Aeolus.

Above: Caspar after returning from his first bombing flight towards Dover/England (1914). As a war pilot, Lieutenant Caspar was the first to fly a Gotha-Hansa Taube and bombs across the English Channel to Dover in October 1914.

Hamburg plant from the group at any time, and he made use of this option when the army administration discharged him from the air force in mid-1916. Within a very short time, he collected a total of 1.5 million marks from Hamburg business and financial circles and then used this capital to found Hanseatische Flugzeugwerke Karl Caspar AG (HFC) on February 19, 1917, with its headquarters in Hamburg-Fuhlsbüttel.

Factory Facility and Airfield:
The company owned two hangars on the (Hamburg) Wandsbek parade ground when it was founded in November 1911, which were destroyed by fire in the summer of 1913. After this fire, the company was relocated to the Fuhlsbüttel airfield, where in 1913, 3 hangars of 300 m² each and an administration building were built, and later another 3 hangars of similar size; 1916 in the summer the flight school burned down, but 6 new hangars with a total of 4,900 m² were built; 1917 in the summer a further expansion of the manufacturing area took place by renting the existing airship hangar and outbuildings with about 10,000 m²; 1917/18 in winter another assembly hall of 2,500 m² was completed.

The airfield had an extension of 800 x 1,000 m.

Fabrication:
The company primarily manufactured new aircraft, but also repaired aircraft and produced spare parts for all types. Hanseatische Flugzeugwerke Karl Caspar AG specialized in particular in the repair of G aircraft.

From the summer of 1917, license production of 200 Albatros C III aircraft began, followed by series production of 75 Friedrichshafen Fdh G.III and G.IIIa aircraft from February 1918.

There were
- 1917   15 repair aircraft per month,
- In 1918   35 new-build C aircraft per month and 5 repair G aircraft and spare parts were produced.

The following were built under license:

Staff and Designers:
The workforce grew from 150 workers in January 1916 to around 1,100 at the beginning of 1918. In the 4th quarter of 1918, the number even rose to almost 1,800.

Besides Ernst Heikel the engineer Hergt was the main designer, who, among other things, designed a single-seater

Aircraft Built Under License by Hanseatische Flugzeugwerke AG (Karl Caspar), Hamburg-Fuhlsbüttel (Hansa)

Type (All Army Aircraft)	Engine [hp], Manufacturer	Crew	Armament	License Built By	Year	Notes
Alb C.III (Hansa)	150, Benz Bz.III 160, Merc. D.III	2	1 MG		1917	100–200 orders
Fdh G.III (Hansa)	2x260, Merc. D.IVa	3	2–3 MG, 800 kg bombs		1918	35 orders
Fdh G.IIIa (Hansa)	2x260, Merc. D.IVa	3	2 MG, 800 kg bombs		1918	40 orders

fighter in 1918.

Aircraft Development:

When the "Hansa- und Brandenburgische Flugzeugwerke AG" was founded, Karl Caspar reserved the right to spin off the Hamburg company again and also made use of this right at the beginning of 1917 when he was again dismissed by the army administration to manage his former company. The "Hanseatische Flugzeugwerke Karl Caspar AG" was now founded.

When in the summer of 1916 an explosion destroyed the zeppelins in the hangar and also the roof of the airship hangar, the Navy decided to abandon the hangar. Caspar had the roof repaired and took over this hangar on a lease basis.

Apart from the two two-seater Hansa Taube (with 50 hp Rumpler "Aeolus" and 95 hp NAG engine respectively) in 1912 and a replica of the Gotha Taube (Gotha-Hansa Taube) with 105 hp Daimler D.I engine, no other pre-war aircraft were manufactured. Between 1917 and 1918, a twin-engine (2 x 110 hp Oberursel UR.II or LeRhone) biplane was built. During a test run, an engine mount broke and fatally injured an assembler. As a result, further development activities were abandoned. Further attention was focused on the licensed production of Albatros C.III reconnaissance aircraft as well as Friedrichshafen Fdh G.III and G.IIIa.

Under the (unconfirmed) designation HFC D1, construction of a 2-engine single-seater fighter developed by Ernst Heinkel was begun before the end of the war. However, Ernst Heinkel was not contractually employed by Caspar for about a year until mid-1921, which again casts doubt on the origin statement. After an engine mount failed during testing and a mechanic was fatally injured, this project was abandoned.

Production Supervisors (Bauaufsicht): #30

Construction Supervision No. 30 was established on January 19, 1917. The aircraft repair shop of the Max Oertz yacht yard and the surface repair and spar fabrication department of the Lüneburg furniture factory were also assigned to this construction supervision. The 14-man supervisory staff included 3 officers and an engineer from Idflieg.

War Material Found by the Inter-Allied Commission of Control (IACC) During Inspections After the End of the War:

75 Airplanes
48 Engines

Situation after WWI:

At the beginning of 1919, the "Hansa" works produced furniture. In July it was sold to a syndicate, which resulted in the cessation of all production.

In 1920 it was taken over by a new company, Caspar Werke GmbH (based in Lübeck-Travemünde). The company was now called Caspar Werke GmbH, Hamburg branch.

By the beginning of 1920, a large part of the halls had been dismantled. The tools that had been used to manufacture the aircraft had been sold.

Note: In September 1918, Caspar acquired from the Dutchman Anthony Fokker his aircraft shipyard on the Priwall in Travemünde, but had to give up the production of aircraft after the end of the war, as the Entente had prohibited the construction of aircraft in the German Reich, and limit himself to the production of other products such as gramophone boxes. Caspar therefore closed his factory in Fuhlsbüttel in 1919 and ceded the premises to the Hamburg Senate for 1,550,000 Mark by April 9, 1920.

On April 8, 1920, the articles of association stipulated that the headquarters of the Caspar group of companies would now be Caspar-Werke mbH in Lübeck-Travemünde. The Hamburg plant was thus downgraded to a branch.

Friedrich Christiansen was appointed technical director, and around 1920/1921 he recruited Ernst Heinkel as a designer. Under the strictest secrecy and bypassing the ban on aircraft construction, the two submarine aircraft U 1 and U 2 were subsequently built on behalf of the USA and Japan, as well as the maritime reconnaissance aircraft S I for Sweden.

Endnote:

1 Karl Caspar (* August 4, 1883 in Netra (Hesse-Nassau); † June 2, 1954 in Frankfurt-Höchst) was a German pilot, aircraft manufacturer and lawyer. He gained notoriety,

Caspar

Above: Hanseatische Flugzeugwerke Karl Caspar, Entrance and Administration.

Above: A type A aircraft in front of the airship hangar in Fuhlsbüttel (manufactured by Centrale für Aviatik in spring 1914).). Since Carl Caspar was called up for military service, further development of this design had to be stopped.

Above: Aircraft delivered for repair (including Han C.III).

Above Right: HFW production hall in Fuhlsbüttel.

Right: -Taube in flight.

Above: Hanseatische Flugzeugwerke Karl Caspar, Entrance and Administration.

Above: After the Imperial Navy abandoned the use of airships, the hangar in Fuhlsbüttel stood empty and could be rented by HFW for aircraft production.

Development of Aircraft Factories

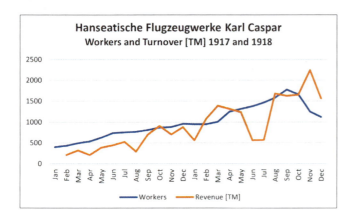

Above: Experimental construction of the twin-engine "HFC", 1918.

particularly in the 1920s, for the types of aircraft developed and built by Caspar-Werke GmbH (formerly Fokker). As an airplane pilot who took his exam before the First World War, he is counted among the Alte Adler (Old Eagles).

Above: View of the airship hangar in Hamburg-Fuhlsbüttel, which was blown up in 1919.

Above & Below: This single-seater fighter was designed in 1918 by Hergt. The aircraft was tested under field conditions without success.

Above & Below: Alb C.III(Hansa)

Above: Hansa flying school.

Above: Experimental construction of the twin-engine "HFC", 1918.

Above: Caspar Taube that bombed Dover.

Above: Fdh G.III(Hansa)

214 Development of Aircraft Factories

4.17 Junkers-Fokker-Werke, AG, Dessau (Junk and Jfa)

Foundation:
Professor Hugo Junkers[1] had been working on aerodynamics since 1908. His laboratory was located in Aachen at the time. In 1911, he built a wind tunnel in Aachen that resembled Mr. Eiffel's.

His numerous experiments led him to conclude that the airplane of the future would be made entirely of metal, with thick wings, no stalks or braces. In April 1915, he had another aerodynamic laboratory with tunnel built at his factory in Dessau, so that he could more easily continue his research and apply the results of his experiments.

This factory was already engaged in metal processing (production of heating stoves, bathtubs, bathtubs, etc.). In September 1915, he had it start building the first German all-metal aircraft (J.1[2]).

The first official tests of this aircraft took place in January 1916. The Flugzeugmeisterei rated them as interesting. The Junkers factory then received its first government order and continued to produce various types of all-metal aircraft.

In the fall of 1917, Junkers merged with Fokker on government orders to increase the production of its factories.

The "Junkers-Fokker-Flugzeugwerke" A. G., which was founded in this way, had new buildings constructed in Dessau. The new company set forth especially the construction of the metal airplanes.

During the war, the Flugzeugmeisterei gave the "Junkers" factories the short name "Junk"[3], after the merger with Fokker: "Jfa"[4]. Fokker near Schwerin had the abbreviation "Jfo"[5].

Factory Facility and Airfield:
Aircraft construction grew from small beginnings in 1914 and developed rapidly in the following years. While the initial space was 200 m², by the summer of 1916 it had grown to 3,800 m² and by the end of World War 1 it had grown to over 20,000 m².

Fabrication:

Staff and Designers:
At the time of its founding in June 1915, the workforce consisted of 10 workers and 4 civil servants; by the end of the war, a total of over 1,200 men were employed.

Prof. Junkers himself acted as designers, as well as Dr. Ing. Mader in particular. The latter also headed Junkers' research institute in Dessau and was instrumental in the development of the world's first all-metal aircraft.

Aircraft Development:
In May 1914, Professor Junkers, who at the time was primarily concerned with the design and construction of his opposed-piston engine, began preliminary work on the construction of a metal monoplane without bracing. Junkers recognized in the braceless cantilever wings a significant advance in the aerodynamic sense, because such a wing has a significantly lower air resistance. For these cantilever wings,

Above: Map of Germany and Dessau.

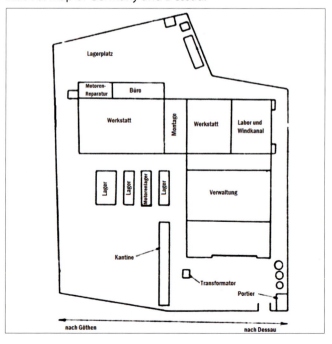

Above: Site map of Junkers at Dessau.

metal was and is definitely preferable to wood because of its uniform strength properties, its weather resistance, its fire resistance and the construction method that is largely ensured with metal.

Following the principles of Eiffel[7] in Auteuil and Prof. Prandtl in Göttingen, Prof. Hugo Junkers created a wind tunnel facility to establish the scientific basis for his design. In May 1914, the first facility was commissioned in Aachen, followed by a second facility in Dessau in February 1916. After the positive laboratory results led to the conclusion that the wing required for the cantilever design with a relatively large airfoil height compared to the usual thin airfoils could be built not only without a noticeable deterioration in the lift-to-drag ratio but with more favorable stability characteristics, construction of test flights could begin. These first tests were carried out in 1915 at the Junkers company

Above: Strength test on a Junkers J.2 prototype under supervision of Bauaufsicht.

Below: Junkers aircraft production (top: J.1, below right: J.2).

Aircraft Built Under License by Hanseatische Flugzeugwerke AG (Karl Caspar), Hamburg-Fuhlsbüttel (Hansa)

Type	Engine [hp], Manufacturer	Crew	Armament	License Built By	Year	Notes
Landplanes						
Junkers J1	120, Merc. D.II	1	-		1915	1 prototype
Junkers J2	160, Merc. D.III	1	-		1915	1 (6)
Junkers J3	-	-	-			Built in parts only
Junk J.I (J4)	200, Benz Bz.IV	2	1-2 MG		1917	227
Junkers J7	160, Merc. D.III	1	-		1917	1 prototype
Junk D.I (J9)	185, Benz Bz.IIIavü				1917	40 Jco 20 Jfa
CL.I (J10) (J.8=prototype)	185, BMW BMW.IIIa				1918	43 (47)
Floatplane						
CS.I (J11)	195, Benz Bz.IIIb	2	3 MG		1918	3 built

in Dessau. Iron was used as the building material and a steel tube construction was used for the wing interior. The wing surface (covering) was designed as a flexurally rigid beam and was thus part of the load-bearing wing structure. The fuselage structure was also made of steel tubes.

In January 1916, the first experimental aircraft, the J1 (monoplane with 120 hp Mercedes D II engine) was demonstrated, followed by an order for 6 aircraft. The rapid development of the aircraft as a weapon on both sides of the front led to a permanent increase in the requirements for flight characteristics and performance. For example, Junkers was forced to make technical changes to the monoplane while the first small order was being processed, among other things to improve the climb performance of the relatively heavy aircraft.

During this period (late 1916) Junkers received a trial order to build 3 experimental J-airplanes (armored biplanes for ground combat) with 200 hp Benz engines. Although the J4 (military J.I) aircraft, which underwent its flight tests at Döberitz in February 1917, had a mass of 2,000 kg, which was enormous by the standards of the time, it proved that the required climb times could also be achieved in metal construction. The first armored aircraft for the front were released in the summer of 1917 and had quite satisfactory results, so that series production could begin. Duralumin instead of steel was used for the first time in the J4/J.I. Due to the lower strength, reinforcing corrugated sheets made of duralumin were also used for the first time in the structural area of the J4. Incidentally, there was also a single J4 in which the entire fuselage, including paneling, was built from duralumin for test purposes.

In parallel development, the single-decker was converted into a D.I and CL.I aircraft, equipped with 185 hp BMW IIIa and 160 hp Mercedes D. IIIa engines, respectively. In September 1917, shortly before the merger with Fokker-Werke, the company produced a single-seat monoplane, followed by a two-seater in December. Comparison flights carried out at Adlershof in the spring of 1918 showed satisfactory results, but the aircraft was never put into larger-scale production. Development work on an R-airplane and a naval D-airplane was equally unsuccessful.

All in all, it can be said that the two partners, Fokker and Junkers, pursued too different approaches to aircraft construction, so that a fruitful collaboration did not materialize after the takeover. It was not until after the end of the First World War, when both entrepreneurs went their separate ways again, that decisive impetus was given to aircraft construction by both companies.

Story behind the fusion from Fokker and Junkers (taken from "The Hugo Junkers Homepage" by Horst Zoeller
The initial Junkers aircraft J1 to J3 and the J4 prototypes were still produced within the Junkers & Co. workshops. Following the successful tests of the Junkers J4, Idflieg was interested in a serial production of this combat aircraft. However, Idflieg was not convinced with the Junkers facilities in Dessau, as Junkers did not have any experience in a large-scale serial production of aircraft. Nevertheless, an initial order for 50 Junkers J.4s was placed in February 1917. Difficulties with the material supply of the armour plates, which were provided by Hüttenwerk Dillingen and change requests from the still ongoing test program at Adlershof resulted in the Idflieg expected delivery delays. The first aircraft 100/17 was delivered in August 1917 for front tests. At this time Junkers was still unable to put out the promised number of 30-50 aircraft per month. Therefore, Idflieg refused to put any further orders for the

Above: From the Junkers experimental aircraft project J.3 only fuselage and wings were built.

Above: The Junkers J.4 (J.I) had a special armored forward fuselage protecting the cockpits and engine.

Above: View at the instrument panel and the two machine guns of a Junkers J.11 (CL.I)

Above: Loading Junkers J.9 (D.I) for the front line.

Above: Junker J.4 (J.I), J.140/17 ready for take-off.

Above: Junker J.4: final completion at Dessau plant.

J4 until Junkers solved his production problems. Idflieg also promoted a merger with one of the larger experienced aircraft manufacturers in Germany. Due to the missing Idflieg orders Junkers got into financial problems and therefore started discussions with several potential partners, like Daimler, Stinnes, Siemens-Schuckert, Bosch, Aviatik, Oesterreichsiche Fiat, and Fokker as well.

The experienced manufacturers were interested in picking up the J4 design for their own production lines and offered Junkers a minority shareholder position. But Hugo Junkers still intended to set up an own production line at his facilities at Dessau and was therefore looking for a partnership with knowledge transfer. Camillo Castiglioni of the Oesterreichische Fiat made a promising offer leaving Hugo Junkers with 70 percent of the shares. The joint venture company was intended to take a license agreement for the J4 from the Forschungsanstalt at Dessau and Fiat would be responsible for the setup of a serial production in the new company. However, these discussions did not come to a successful end.

Above: Junkers J.4 (J.I) J.884/17 at airfield in Köln-Butzweilerhof (probably postwar).

Above: Demilitarized Junkers J.10 with cabin configuration for civil air transport, here flying between Dessau and Weimar, where the National Assembly of the Weimar Republic met in 1919.

Anthony Fokker was another discussion partner. Instead of picking up the J4 for his own production lines, he offered the setup of a new facility at Dessau. As the Fiat offer, the Fokker offer was heading for a license agreement between the new company and the Forschungsanstalt. Junkers and Fokker should hold 50 percent of the company. On 20th October 1917 the Junkers-Fokker-Werke A.G. was founded at Dessau with a capital of 2.6 Mio. Mark. The Junkers capital was added with facility grounds, tools and materials

Following the foundation of the Junkers-Fokker-Werke Idflieg promptly put several orders for Junkers aircraft. With the engagement of Fokker at the Junkers facility and with additional staff from the Fokker-Werke at Schwerin the production rate at Dessau was raised from 17 J4 aircraft in January 1918 to 31 aircraft in October 1918. However, this was still less than the promised output rate of 50 aircraft per month. Initially the J.4s were shipped to Döberitz for flight tests, still in 1917 the flight tests were transferred to Wurzen near Leipzig. In May 1918 Junkers was able to offer test flights at a newly established airfield at Dessau as well. At least with the production setup of the Junkers J.9 new complexities arise.

To increase the number of available J4 aircraft, in Summer 1918 Idflieg placed orders at other aircraft manufacturers for the J4 as well. In June 1918 a number of 100 J4 aircraft were ordered at Linke-Hofmann in Breslau. But as the license discussions between Linke-Hofmann and the Junkers did not make progress, this order was later cancelled. Another third-party order was put by Idflieg at Hansa-Brandenburg for a total of 50 Junkers J9 aircraft. Again, this third-party order was not realized. Therefore, all WWI Junkers aircraft were produced at Junkers & Co. or at Junkers-Fokker-Werke. At the end of WWI deep differences between Junkers and Fokker existed. During the time of Junkers-Fokker-Werke Fokker always tried to transfer technological knowhow from the Junkers-Fokker-Werke to his own facilities at Schwerin. Additionally, Fokker took opposite positions to Junkers' designs as he liked to promote his own aircraft at Idflieg. On the other hand, Jco and Jfa were competitors regarding Idflieg orders, at least a number of pre series aircraft were built at Jco. Therefore, when WWI came to an end, both Hugo Junkers and Anthony Fokker quickly separated from each other, and the Junkers-Fokker-Werke were dissolved.

Production Supervisors (Bauaufsicht): #36

This construction supervision of the Air Force Inspectorate was relatively small compared to other sites. It consisted of one officer, one post-flight pilot and another seven men.

War Material Found by the Inter-Allied Commission of Control (IACC) During Inspections After the End of the War:

6 airplanes
1 engine.

Situation after WWI:

After the armistice, the "Junkers-Fokker Flugzeugwerke A. G." was closed (April 21, 1919) Anthony Fokker left Germany and left his shares in Junkers-Fokker AG to Hugo Junkers on December 3, 1918, so that he regained independence over his company. In June 1919, Junkers-Fokker AG was renamed „Junkers Flugzeug-Werke AG." By the end of 1920, Junker-Flugzeug-Werke was mass-producing F13 100 hp limousine aircraft. The company also produced industrial engines and various fittings for bathtubs and water heaters.

Endnotes

1 Heinrich Hugo Junkers (* February 3, 1859 in Rheydt; † February 3, 1935 in Gauting) was a German engineer and industrialist. He founded Junkers & Co. in Dessau in 1895 and owned Junkers Motorenbau GmbH and Junkers Flugzeugwerk AG until 1932. Initially known as a designer of gas heaters, Junkers developed fundamental innovations in aircraft construction, such as all-metal construction and the corrugated structure, as a university teacher and researcher, engineer and entrepreneur. In addition, his group also experimented with aircraft engines, which became especially popular after the end of World War I. He founded the airline Junkers Luftverkehr AG, one of the forerunners of Luft Hansa.

2 Not to be confused with the later J.I (factory designation J4).

Junkers

Above: Junkers J.1 (1915)

Above: Junkers J.2 (1915)

Above: Junkers J.3 (only fuselage has been built, not a complete aircraft) (1916)

Above: Junkers J.I (Junkers J.4) (1917)

Above: Junkers J.7 (1917)

Above: Junkers D.I (Junkers J.9) (1917)

Above: Junkers CL.I (Junkers J.10) (1918)

Above: Junkers CLS.I (Junkers J.11) (1918)

3 Junk - Junkers
4 Jco – Junkers & Co., Dessau (bis 19. Oktober 1917)
5 Jfa – Junkers-Fokker A.G., Dessau (ab 20. Oktober 1917
6 also referred to as „Marine C" or „CLS.I"
7 Alexandre Gustave Eiffel (b. December 15, 1832 as Alexandre Gustave Bonickhausen called Eiffel in Dijon; † December 27, 1923 in Paris) was a French engineer and designer of steel structures, including the Eiffel Tower in Paris and the Statue of Liberty in New York.
8 In 1909, Gustave Eiffel built a laboratory with a wind tunnel on the Champ de Mars, which was moved to Auteuil in 1912.
9 Ludwig Prandtl (* February 4, 1875 in Freising; † August 15, 1953 in Göttingen) was a German engineer. He made significant contributions to the fundamental understanding of fluid mechanics and developed boundary layer theory.

Development of Aircraft Factories

4.18 Kaiserliche Werften

Foundation:
The Imperial Shipyards in the German Empire were state-owned shipyards, alongside private shipyards, responsible for the design, construction, repair, modernization, and conversion of warships and submarines of the Imperial Navy in the period from 1871 to 1920. Some of their history began as Royal Shipyards for the Prussian Navy and, from July 7, 1867, for the Navy of the North German Confederation.
In addition to the 3 shipyards in the German Empire:
Kaiserliche Werft Danzig (1852-1920)
Kaiserliche Werft Kiel (1867-1920)
Kaiserliche Werft Wilhelmshaven (1853-1920)
the Kaiserliche Werft Antwerpen in Hoboken (1914-1918) continued to exist at the time of WW1.
The Ottoman Navy also had facilities in Istanbul, İzmit and Gemlik known as Tersâne-i Âmire (Turkish for Imperial Shipyards).

Repairs were carried out at the Kaiserliche Werft, as well as the construction of licensed and own designs. In addition, a whole series of experiments were carried out in the interest of naval aviation. At Danzig-Putzig, Coulmann[1] carried out experiments with central floats around 1912, using a converted Albatros fuselage biplane. Around 1917/18, the first catapult starts of a Hansa-Brandenburg from a military ship also took place here.

Above: Map of Germany with Kaiserliche Werft locations.

Surprisingly, in addition to the already established aircraft companies, the Navy administration decided to develop and build aircraft on its own responsibility from the end of 1914. All 3 above-mentioned shipyards of the Kaiserliche Marine built a total of only 20 floatplanes, a number that seems insignificant when measured against the total number of all aircraft in service with the Navy.

Kaiserliche Werft Danzig-Putzig (now Gdansk, Poland)

Factory facility and airfield:
The location of the shipyard is shown in the adjacent figure. Aircraft had access to the Vistula River (Weichsel) via a slipway, where takeoffs and landings took place.

Fabrication:
During World War I seaplanes were built, maintained and repaired on a small number for the Kaiserliche Marine just as in the other Imperial shipyards.
Marine No.: 404, 405: KW Type 401
Marine No.: 467 – 470: KW Type 467
Marine No.: 1105–1106: KW Type 1106
Marine No.: 1650: KW Type 1650

Aircraft Development:
Marine numbers 404 and 405 were the sole two examples of a unique seaplane design produced for the flying service of the Imperial German Navy during the First World War. By 1917, the output of the major German seaplane manufacturers was taken up producing machines for front-line service. As a consequence, the only machines available for training purposes were those that had been made obsolete or which had been damaged and rebuilt. In order to provide modern trainers for the Navy, the Kaiserliche Werft Danzig undertook the design and construction of two brand-new seaplanes between March and June, unarmed two-seat biplanes. These machines were supplied to the naval base at Putzig along with a batch of four trainers of a different design, numbered 467–470.

Construction of these unarmed two-seat biplanes took place between October 1916 and March 1917., ahead of a separate order for two more machines of different design that had been assigned lower serial numbers by the Navy (404–405).

\multicolumn{7}{c}{Aircraft Built by Kaiserliche Werften Danzig}						
Type	Engine [hp], Manufacturer	Crew	Armament	License Built By	Year	Notes
KW Type 401	100, Merc. D.I	2	-		1917	2, 404, 405
KW Type 467	150, Benz Bz.III	2	-		1916/17	4, 467-470
KW Type 1106	150, Benz Bz.III	2	-/1 MG		1917	2, 1105, 1106
KW Type 1650	220, Merc. D.IV	2	1 MG		1917	1

Kaiserliche Werften

Above: Stamp of Kaiserliche Werft Danzig.

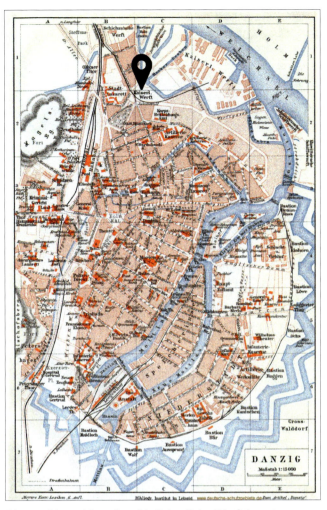

Above: Map of Danzig with Kaiserliche Werft location.

Imperial German Navy seaplanes numbers 1105 and 1106 were the only examples of a unique design produced for the navy's flying service (not for training purposes) during the First World War. They were unarmed biplanes of conventional configuration with staggered wings of unequal span. The empennage included a sizable ventral fin. Intended as training aircraft, the pilot and instructor sat in tandem, open cockpits. The undercarriage consisted of twin pontoons. The interplane strut arrangement was remarkable for its day, consisting of N-struts and V-struts without any rigging wires.

These machines were supplied also to the naval base at Putzig at the end of 1917. After a short time at least the 1105 were returned to the manufacturer as "unserviceable".

Number 1650 was an armed reconnaissance seaplane equipped with radio equipment capable of transmission and reception, therefore gaining the naval CHFT classification.

Situation after WWI:

After the separation of Danzig (Gdansk) from the territory of the Reich after the end of the war, it continued to operate as the Gdansk Shipyard and Railway Workshops from 1919.

Kaiserliche Werft Danzig employed around 7,000 workers at the end of World War I and became a municipal enterprise under the leadership of the victorious powers in the post-war period. With 30 percent share each of Great Britain and France and 20 percent shares each of Poland and Gdansk, The International Shipbuilding and Engineering Company Limited - Gdansk Shipyard was established. In November 1919 Ludwig Noé took over the management of the shipyard, which still had 2500 employees.

Above: The aircraft with the marine designations 1105 and 1106 were of the same design (type 1106).

Above: This seaplane is a mystery. According to the Navy Directory, this was a Brandenburg Type W.1 built at KW in Danzig. Other sources (J. Herris in [*Albatros Aircraft of WWI, Vol. 3,*]) refer to this seaplane as the Albatros W.1.

Development of Aircraft Factories

Kaiserliche Werft Wilhelmshaven

Above: Stamp of Kaiserliche Werft Wilhelmshaven.

Above: Map of Kaiserliche Werft Wilhelmshaven

Above: Entry gate of Kaiserliche Werft in Wilhelmshaven.

Foundation:
Shortly after the commissioning of the war port in the middle of the 19th century, the construction of the third royal Prussian shipyard in Germany was started on the site in 1870, following the already existing royal Prussian shipyards in Danzig and Kiel. With the proclamation of the German Empire in January 1871, the navies of the North German Confederation and Prussia were in turn combined to form the Imperial Navy, and the former "Königliche Werften" (Royal Shipyards) were renamed "Kaiserliche Werften" (Imperial Shipyards) accordingly.

In the meantime, a new town had grown up on the area around the naval facilities, which was given the name Wilhelmshaven in 1869 on the occasion of the inauguration of new harbor facilities by Wilhelm I.

The company did not turn to aircraft construction until the end of 1914. For this purpose, vacant hangars on the shipyard's waterside were used.

Fabrikation:
During World War I, the Wilhelmshaven shipyard, like the other two Imperial shipyards, manufactured seaplanes exclusively for the Imperial Navy. The identification numbers of the aircraft were:
401-403: KW Type 401
461, 462: W10 and KW Type 462 respectively
945: KW Type 945
947: KW Type 947.

Aircraft Development:
Imperial German Navy seaplanes numbers 401 to 403 were the only three examples of a unique seaplane design

Aircraft Built by Kaiserliche Werften Wilhelmshaven						
Type	Engine [hp], Manufacturer	Crew	Armament	License Built By	Year	Notes
KW Type 401	100, Merc. D.I	2	-		1915	2, 404, 405
KW Type 462	150, Benz Bz.III	2	-		1916/17	4, 467-470
KW Type 945	150, Benz Bz.III	2	2 (3) MG		1917	2, 1105, 1106
KW Type 947	220, Merc. D.IV	2	1 MG		1917	1
W10 (#461)	150, Benz Bz.III	2			1916	

Kaiserliche Werften

Above: View thru the fuselage of Marine Number 401.

Right: Bad end of Marine Number 401, built at Kaiserliche Werft in Wilhelmshaven.

Below: Sea reconnaissance aircraft 947 with Daimler D.IV engine.

produced for the Navy's flying service during the First World War. Production of these types commenced in April 1915 in an effort to supply the navy with a seaplane trainer of contemporary design. With the outbreak of war, the output of Germany's major seaplane manufacturers was taken up with producing front-line types, and the only trainers available were obsolete or rebuilt machines withdrawn from their original duties. Number 401 and its two siblings were delivered to the Navy in August 1915.

The KW Wilhelmshaven Marine Numbers 461 and 462 (Type 461, also known as W 10) were of the unarmed, two-seat floatplane (B-type) genus according to the Kaiserliche Marine's aircraft group classification.

Marine number 945 (also called W9) was based on the Brandenburg W12, completed in the summer of 1918, but almost certainly never delivered to the Navy.

Marine number 947 was handed over to the SVK in December 1917, but the aircraft was not accepted and returned to the shipyard.

Situation after WWI:

After World War I, the shipyard was operated by the Reichsmarine (from 1935: Kriegsmarine). Since 1957, the former shipyard site has been home to a naval arsenal of the German Navy (since 1990: Deutsche Marine).

Kaiserliche Werft Kiel

Above: Stamp of Kaiserliche Werft Kiel.

Foundation:
When the Kaiser Wilhelm Canal was finally completed in 1895, the Kaiserliche Werft Kiel moved to Kiel-Gaarden-Ost in 1899. Between 1899 and 1904, the area of the shipyard expanded to such an extent that the Germania shipyard in the south had to cede part of its site to the Kaiserliche Werft. The operation also continued to grow to the north, and in 1904 the last remnants of the old fishing village of Ellerbek disappeared with the expansion of the shipyard. To connect the two parts of the shipyard, the suspension ferry was completed in 1910 and soon became a landmark in Kiel.

Fabrication:
Forever reason, the Kaiserliche Werft Kiel took part in the aircraft production without any experience in this business. Two orders of each 2 aircraft type KW 462 were given in 1915.

The four Type KW 462 seaplanes to be built at the Kaiserliche Werft Kiel were to be identified as 463 to 466.

It is questionable whether all four aircraft have been delivered. The only certainty is that Navy number 463 was delivered to the Navy in mid-1917 as a two-seat unarmed trainer aircraft.

Aircraft Development:
Kaiserliche Werft Kiel received an order for four seaplanes in October 1915. The first aircraft of this batch (# 463) was

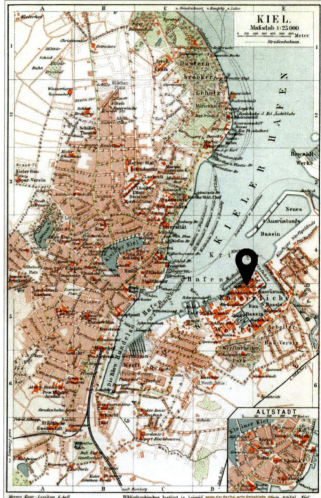

Above: Map of Kaiserliche Werft Kiel.

Above: Kaiserliche Werft in Kiel.

delivered to the seaplane testing unit (SVK – Seeflugzeug-Versuchskommando) at Warnemünde near Rostock the

\multicolumn{7}{c}{Aircraft Built by Kaiserliche Werften Kiel}
Type	Engine [hp], Manufacturer	Crew	Armament	License Built By	Year	Notes
KW Type 462	150, Benz Bz.III	2	-		1916	1 (4), 463-466

Kaiserliche Werften 225

Above: The only seaplane built in Kiel was Marine number 463 (Type 462). The ordered #464 – 466 were not finished.

aircraft being used as a trainer at Warnemünde itself.

Marine number 463 was a conventional, two-bay biplane with unstaggered wings of equal span and two open cockpits for the pilot and instructor. The undercarriage consisted of twin pontoons. The large, square rudder was hinged to the rear end of the fuselage and extended below the ventral line of the fuselage.

Situation after WWI:
After the end of the First World War and the final liquidation of the Kaiserliche Werften in 1920, the Kaiserliche Werft Kiel initially became the Reichswerft Kiel. From 1925, it was part of Deutsche Werke AG in Berlin as Deutsche Werke Kiel AG (DWK).

Endnote

1 Wilhelm Coulmann * 13.08.1880. Royal Navy Shipwright Master, Pilot License Number 269.

following summer. It remained there under test for well over one year and was not finally accepted for service until summer 1918. This lengthy delay was possibly due to the

Above: KW Type 401 (Danzig) (1917)

Above: KW Type 401 MN 404 (Danzig) (1917)

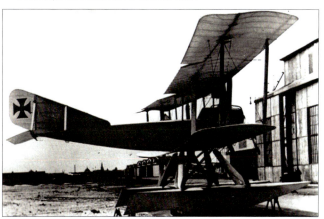

Above: KW Type 462 (Wilhelmshaven, Kiel) (1916)

KW Type 945 (Wilhelmshaven) (1917)

KW Type 947 (Wilhelmshaven) (1917)

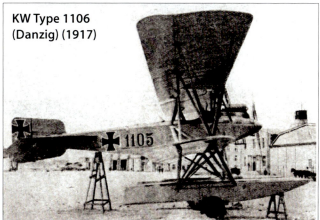

KW Type 1106 (Danzig) (1917)

Development of Aircraft Factories

Hansa-Brandenburg model "W" built at KW in Danzig.

Kiel, Kaiserliche Werft

Kiel, Kaiserliche Werft

Kaiserliche Werft Wilhelmshaven, 1894

Kaiserliche Werft #462

Kaiserliche Werft #461

Sea reconnaissance aircraft 947 with Daimler D.IV engine.

4.19 Kondor-Flugzeugwerke, Essen (Kon)

Above: Advertisement 1916.

Foundation:
Kondor-Flugzeugwerke, founded by the Lord Mayor of Düsseldorf, Wilhelm Marr, in Essen an der Ruhr on July 20, 1912, maintained its main plant on the site of the former Gelsenkirchen-Essen-Rotthausen airfield. The company also maintained a flying school there. A second flying school was opened in Nordhausen.

Factory Facility and Airfield:
The factory site in Essen covered approximately 1,200 m² at the time of its foundation, which developed rapidly. In 1917, at the peak of production, the site area was 15,000 m². The Kondor aircraft works together with the flight school in Nordhausen covered a terrain of around 1 million m².

Additional workshops were purchased in August 1918 from Kondor-Gesellschaft in Lemgo for woodworking needed for the construction of Kondor aircraft in the Essen factory. They covered an area of 10,000 m². In 1918 they employed 20 workers.

Fabrication:
Shortly after its foundation in 1912, the company launched its own aircraft designs and achieved success with the Tauben they designed and built.

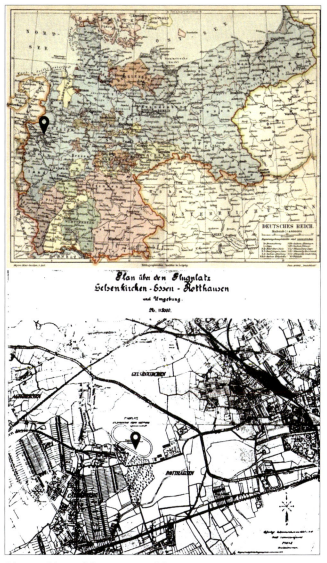

Above: Map of Germany and Essen.

Kondor site plan

Several versions and types of the "Kondor" were subsequently developed, including a small machine Kondor D.1, also designated as "Kondor-Laus" (Kondor Flea). But success was not forthcoming.

Despite the good results, no machines could be sold. The shareholders continued to invest money in the company, but it could no longer be sustained. On March 30, 1914, it was decided to liquidate the company. But before this could happen, construction orders were received from Spain

Development of Aircraft Factories

Aircraft Built by Kondor-Flugzeugwerke, Essen (Kon)							
Type	Engine [hp], Manufacturer	Crew	Arma-ment	License Built By	Year	Notes
Monoplane*	100, Argus	2				
Tauben Type A* - H2*	100-120, Argus or Merc.	2				26
Kondor biplane Type A*, 2*	120, Merc. D.II	2				4
Kon B.I	?	2				1
Kon W.I³*	120, Merc. D.II	2				prototype
Kon W.IIC³*	220, Merc. D.IV	2				prototype
Kon W.III						prototype
Kon W.IV						prototype
Kon Dr.I	160, Merc. D.IIIa	1				prototype
Kon D.I⁴*	110, Oberursel UR.II	1				1 (#200)
Kon D.II⁴*	110, Oberursel UR.II	1				2 (#201)
Kon D.6⁴*	110, Oberursel UR.II	1				1
Kon D.7⁴*	160, Merc. D.III	1				1
Kon E.III	160, Oberursel UR.III	1				<10 (100)
Kon E.IIIa	160, Goebel Goe.III	1				
Alb B.II(Kon) Alb B.IIa(Kon)	100-120, Argus, Benz, Merc.	2				350

Designed by: *Suwelack, 2*Beck, 3*Westphal (therefore the "W"), 4* Rethel.
At the time of the armistice, the E.IIIa was considered one of the best "D" aircraft in Germany.

and the company continued to build. Due to the outbreak of war in 1914, however, the aircraft was not delivered. After a demonstration flight in Cologne, the "Kondor" was purchased immediately, and the other aircraft destined for Spain were also delivered to the army administration.

In 1915, due to good experience with this Spanish G export version, the factory received an Army order for several Type H "Kondors" with 100 hp Daimler D.I engine.

In 1915/16 the company was engaged in building repairs, and in 1917/18 in the production of training aircraft. A total of about 480 aircraft left the Rotthausen Kondor plant.

Staff and Designers:
Due to the army orders, the Kondor-Werke operation could be significantly expanded. The workforce reached a strength of 1,200 men in the summer of 1917.

The construction of aircraft could then be resumed. Responsible for this were:
Josef Suwelack[1]
Paul Westphal and
Walter Rethel[2]

Aircraft Development:
Although Suwelak, as the most experienced of the Kondor pilots, demonstrated the first pigeons designed by the company in flight, at first no orders were placed for the pigeons. The company endeavored to produce aircraft types through further improvements, which were not entirely without success in aircraft competitions and records, such as:

1913: 8-hour endurance record, Prinz Heinrich flight, Aeroplan tournament in Gotha, flight around Berlin, Krupp flight week, triangle flight;

1915: Altitude record 4,000 m with 4 passengers (Lieutenant Höhndorf[5]).

Since the German army administration could not decide on the orders in the Kondor-Flugzeugwerke, the company first turned to the foreign market and exhibited its own design of an aircraft in Madrid on February 1, 1914. This exhibition and demonstration was a complete success, as the Kondor-Flugzeugwerke received a trial order from the Spanish government for the delivery of a series of Kondor Taube G. This seemed to secure the further development of the company, but as a result of the outbreak of war, the delivery of the aircraft was prohibited by the army administration, as the aircraft were needed by Germany's own aviation forces. Due to the difficulties associated with the outbreak of war, such as the mobilization of a large part of the personnel, including the technical director at the time, Mr. Josef Suwelack, the company itself ran into difficulties, although it was itself constantly seeking Army orders. By

Above: Kondor-Works in Essen-Rotthausen.

Left: Kondor-Werke 1913: Wooden wing production for Tauben.

Above: February 26, 1913: Suwelak's Kondor-Taube in Deventer near Zwolle. He attempted to fly to London. The bad weather conditions forced him to abort the flight.

Above: Kondor E.IIIa prototype, aft view.

Right: Kondor advertisement in *Motor*, 1913.

Above: Kondor W.1 powered by a Daimler D.II engine. In 1915 reached Höhndorf as pilot together with 4 passengers a new high-altitude record (4.000 m).

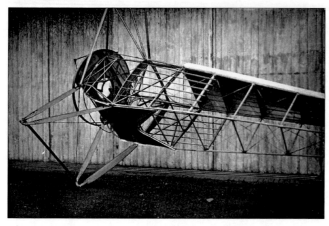

Above: Kondor E.III: Tubular steel construction of the fuselage structure.

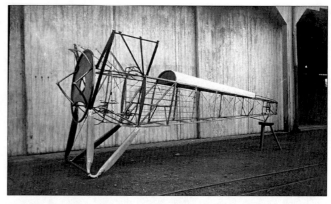

Above: Wooden landing gear on the Kondor E.III.

Above: Kondor E.III: 4-spar wing structure.

hiring new designers, 4 biplanes D.I, D.II, D.6 and D.7 were developed in succession.

In 1915, due to good experience with this Spanish G export version, the factory received an Army order for several Type H "Kondors" with 100 hp Daimler D.I engines.

In 1915/16, they began to design various school biplanes, which had angular fuselages as opposed to the round fuselages of the machines they had previously built.

Of these, only the Kondor W.1 and Kondor W.2c types have so far become known in the picture. In 1915, the company has built Albatros biplanes under license, and for 1916 it had received an order from the Army Administration for 50 Albatros type B.II training aeroplanes.

First a triplane with a 160 hp Mercedes D.IIIa engine was built and then a two-seater unit training aircraft with a 120 hp Mercedes engine. However, the army administration continued to give preference to the Albatros aircraft.

In the meantime, a subsidiary company had been established in Nordhausen/Harz, where the Kondor flight school was located, which was called "Kondor-Werk, Gesellschaft für Holzbearbeitung mbH". Another company division, "Kondor-Flugzeugwerke GmbH Abt. Sägewerk" in Lemgo/Lippe, was added in 1918 by taking over the Schnackenback sawmill, because good wood and its good processing were essential for aircraft construction. And a lot of it was needed.

While previously only training aircraft had been built, the first fighter aircraft was built in 1918, the Kondor D.I. It was a normal braced biplane with a round plywood fuselage and V-stems. It was equipped with a 110 HP Oberursel UR.II engine and took part in the second comparison flight in June 1918 in Berlin-Adlershof.

The Kondor D.II also took part in the same test flight. It was an improved version of the Kondor D.I. The engine remained the same. The V-struts were replaced by II-struts. An interesting feature of the Kondor D.II was that the ailerons were not only on the upper wing, but also on the lower wing.

The Kondor D.VI was a further development of the Kondor D.II. The engine was an Oberursel UR.III with 145 hp. The special and unusual feature of this aircraft was the separated upper wing to improve upward visibility. No doubt this was ideal for a biplane, but the turbulence at the separation points above the fuselage also brought aerodynamic disadvantages. Incidentally, the fuselage was not made of plywood, as was the case with its predecessors, but consisted of a tubular steel structure with fabric covering.

Another aircraft, the Kondor D.7, had a completely different appearance from its predecessor. With its torpedo shape, it strongly resembled the Albatros aircraft.

It was a biplane with a carefully covered 160 hp Mercedes D.III engine. A striking feature was the narrow lower sweeping wing, which was braced by three struts that diverged from a point to the upper wing. The landing gear also had three shrouded struts on each side of the wheel. The Kondor D.VII crashed during testing at Adlershof in 1918.

Since the airfield in Essen-Gelsenkirchen-Rotthausen was unsuitable for effective training operations, the company transferred its flight school to the military airfield in Großenhain/Saxony on April 20, 1915, and finally began training operations in Nordhausen on September 1, 1917. The training aircraft, mainly Kondor Tauben and Kondor B.I, had the identifiers K1 to K23, but DFW, Albatros and LVG aircraft were also used for training.

Kondor

Right: Aircraft production line. Final assembly of Albatros B.II.

Right: WooWoodworking. Manufacture of the chassis struts.

Additional workshops in Lemgo were purchased in August 1918 by Kondor-Gesellschaft for woodworking needed for the construction of Kondor aircraft at the Essen plant.

They covered an area of 10,000 m². In 1918 they employed 20 workers. At the beginning of 1921, 120 workers produced doors, windows and ammunition boxes.

After the end of the war (until June 1, 1020) furniture was also produced in the "Kondor-Werke". On June 1, 1020, due to bad business, the company stopped production.

Production Supervisors (Bauaufsicht): #23
The production supervision No. 23, established on December 14, 1916, was led by an officer. In addition, there were another 10 men who maintained the service.

War Material Found by the Inter-Allied Commission of Control (IACC) During Inspections After the End of the War:
In Essen: 8 different airplanes, disassembled and in poor condition.
In Nordhausen: 2 airplanes and 3 engines as well as spare parts.

Situation after WWI:
In Essen: In 1919 and until June 1, 1920, the factory "Kondor" produced furniture. On June 1, 1920, ceased production due to poor business.

In Nordhausen: Since 1919, the facility no longer worked for the aviation industry. In 1920, part of the buildings was sold to a company that manufactured water pumps for steam engines. The other part continued to belong to "Kondor Werke GmbH," which in early 1921 primarily manufactured doors, windows and furniture.

In Lemgo: At the beginning of 1921, 120 workers manufactured doors, windows, and ammunition boxes.

After the First World War, aircraft manufacturers lacked customers. The military had ceased to be a customer, civil aviation was still in its infancy and the great enthusiasm for sport aviation had faded. And so the aircraft factory in Rotthausen closed in 1920, but the Kondor company

Right: Wooden Company logo.

continued to exist until 1993 and produced furniture at its site in East Westphalia – not such a big change as it seems from today's perspective, because wood was an important material in the early aircraft models. For example, the wooden propellers for numerous Kondor aircraft came from the company's own plant in Lemgo. Last but not least, Kondor was in good company, because numerous companies that had started out in the young aviation industry soon had to switch.

The "Bund-Deutscher Flieger e. V.", founded in Essen by former wartime aviators, organized the first German flying day after the war on September 21 and 22, 1919, at the Gelsenkirchen - Essen - Rotthausen airport, at which a Kondor E.IIIa-Parasol also took part.

Development of Aircraft Factories

Above: Kondor Taube Type A in flight. (1912)

Above: Kondor Taube Type C. (1913)

Above: Kondor Taube Type E. (1913)

Above: Kondor Taube Type G. (1914)

Above: Kondor Taube Type B. (1913)

Above: Kondor Taube Type D. (1913)

Above: Kondor Taube Type H. (1914)

Above: Kondor Type A biplane. (1914/1915)

Kondor

Above: Kondor W.I with Daimler D.II. (1914/1915)

Above: Kondor W.I with Daimler D.II. (1915)

Above, Above Right, & Below: Kondor B.I, 120 hp Mercedes D.II ("B.I" – internal company designation). (1917)

Above: Kondor W.IIC (1916)

Above: Kondor D.I (1918)

Above & Below: Kondor D.VI (1918)

Above: Kondor D.II (1918)

Development of Aircraft Factories

Above: Kondor D.VII (1918)

Above: Kondor E.III, 160 hp Oberursel Ur.III. (1918)

Above: Kondor E.III, 160 hp Oberursel Ur.III. (1918)

Above: Kondor E.IIIa, 160 hp Goebel Goe.III (1918).

Above: Kondor E.IIIa (1918). Competing at the 3rd Fighter Competition in October, the Kondor E.III demonstrated higher wing strength and roll rate than the Fokker D.VIII. Unlike the Fokker, the Kondor's wing did not vibrate at high speed. The E.IIIa was faster and had a much better climb rate than the E.III due to its engine.

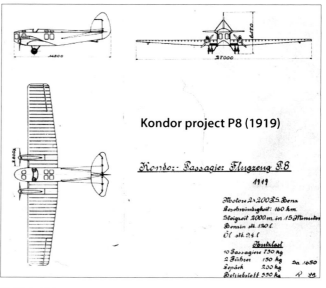
Kondor project P8 (1919)

Above & Right: Albatros B.IIa(Kon) (1918)

4.20 Linke-Hofmann Werke AG, Breslau (now Wroclaw), (Li)

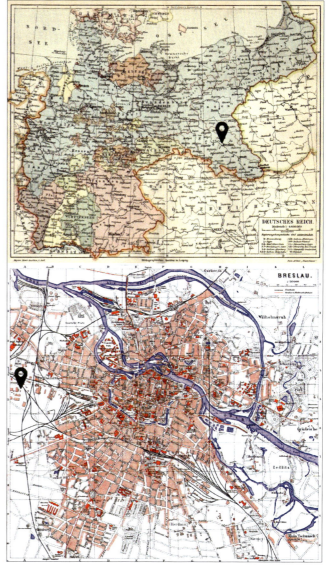

Above: Map of Germany and Breslau.

Foundation:
The original company was founded in 1841 in Breslau by Gottfried Linke and became a joint stock company in 1871. Aircraft construction itself was not started until after the outbreak of World War 1 in 1916.

Factory Facility and Airfield:
The factory underwent major structural expansions on the site in Klein-Mochbern (now Mucho or Mały, Poland) in 1898/99 and 1909 to 1911. At the end of the war, the total area owned by the company was about 1.5 million square meters and had a built-up area of about 175,000 m². A factory building with a built-up area of 11,000 m² and an airfield of about 352,000 m² were available for aircraft construction, adjacent to the factory. This airfield was used for take-off and landings of C-airplanes, while R-airplanes were flown in at the airfield north-east of Wroclaw in Hundsfeld (today Wrocław-Psie Pole).

The production of the aircraft was divided into a total of 12 departments:
1. woodworking, machine shop,
2. part carpentry,
3. locksmith/plumbing shop,
4. wing construction,
5. paint shop,
6. upholstery workshop,
7. fuselage construction,
8. small aircraft assembly,
9. R-aircraft assembly,

Above: Linke-Hofmann production hall in Breslau.

Aircraft Built by Linke-Hofmann Werke AG, Breslau (now Wroclaw), (Li)

Type	Engine [hp], Manufacturer	Crew	Armament	License Built By	Year	Notes
Rol C.IIa(Li)	160, Merc. D.III	2	2 MG		1916/17	1+52
Alb C.Id(Li) & If(Li)	160, Merc. D.III	2				100 ordered
Alb C.III(Li)	160, Merc. D.III	2	2 MG		1917	75
Alb C.X(Li)	260, Merc. D.IVa	2			1917	50
Alb B.IIa(Li)	100–120, various	2			1917/18	~150 (200)
Alb C.XII(Li)	260, Merc. D.IVa	2			1918	25 (order cancelled)
Li R.I (8/15, 40/16)	4x260, Merc. D.IVa	(6)			1916	2 (4) 4 engines central in fuselage, 2 propellers
Li R.II (55/17)	4x260, Merc. D.IVa	(6)	(3 MG, 700 kg bombs)		1918	1 (2), 4 engines central in fuselage, 1 propeller

10. repair department,
11. small aircraft flight test department (Gräbschen airfield),
12. R-aircraft flight test department (Hundsfeld airfield).

For the purpose of quality assurance of production, a total of 4 additional control stations and a material testing station were established. The modern equipment with machine tools for wood and metal processing allowed a largely independent production. Only the instruments, engines and propellers were excluded from in-house production.

Fabrication:
The company manufactured these licensed aircraft:
LFG Roland C IIa,
Albatros B IIa,
Albatros C III,
Albatros C X,

Linke-Hofmann designed and built two R-airplane types; two of the type Li. R.I and one R.II were manufactured.

The Linke-Hofmann R.II was still undergoing flight tests at the end of the war.

Staff and Designers:
In 1916, the company had about 450 employees and 18 civil servants in the aircraft construction department. After the course of a year, this number had risen to about 530 workers with 47 civil servants. In the last year of the war, the number of employees was reduced to 350 workers and 47 civil servants. In addition, 130 workers were employed in R-aircraft construction.

Aircraft Development:
In 1916, Linke-Hofmann Werke were particularly involved in aircraft repairs. At the end of 1916, the company began building new aircraft under license. At the same time, Linke-Hofmann also developed its own designs, especially R-airplanes. The only type built and accepted was an R airplane designated Li R I. A second R airplane design, Li R.II, was undergoing flight tests at the end of the war. A total of 319 aircraft were manufactured in Linke-Hofmann factories.

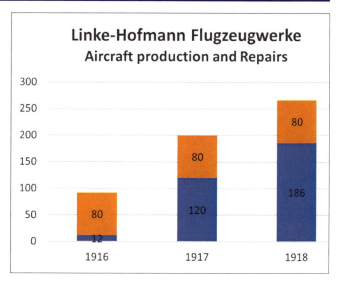

Production Supervisors (Bauaufsicht): #18
The Production Supervision Department, which was responsible for the acceptance of aircraft, the inspection of construction parts and production processes, was established as early as 1916. It was also responsible for the timely procurement of materials of war-important goods and semi-finished products. It was headed by an officer. Subordinate to this officer was an engineer from Idflieg as well as service personnel.

War Material Found by the Inter-Allied Commission of Control (IACC) During Inspections After the End of the War:
2 Giants R.I. und R.II. with engines installed.
105 aircraft Albatros and LVG (repairs) without engines.

Linke-Hofmann

Situation after WWI:
Since 1919, the Linke and Hofmann factory no longer built aircraft equipment. At the beginning of 1921, 8,000 workers were employed in the production of locomotives, wagons, and boilers.

Above: Alb C.X(Li) before acceptance. The Alb D.II standing in front was not part of the production range.

Above: Linke-Hofmann R.I (R40/16) (1917)

Above: Rol C.IIa(Li) (1916/17)

Above: Alb C.III(Li) (1917).

Above: Linke-Hofmann R.I (R.40/16) in front of a hangar in Breslau.

Above: Linke-Hofmann R.II (R.55/17) with 4 Daimler D.IVa engines in the fuselage, one tractor propeller at the nose. The first flight took place in 1919.

Above: Linke-Hofmann R.II (R55/17) (1918/1919)

Above: Alb B.IIa(Li) (1917/18)

4.21 Luft-Fahrzeug-Gesellschaft mbH, Berlin-Charlottenburg (Rol)

Foundation:

Luftfahrzeug Gesellschaft mbH (LFG) is one of the oldest German plants established for the factory production of aircraft. It had its origins in Motorluftschiff-Studiengesellschaft mbH, which was founded in Bitterfeld in 1906. LFG was founded when Motorluftschiff-Studiengesellschaft merged with Parseval-Gesellschaft to build airships in 1908. The company headquarters moved from Bitterfeld to Berlin-Charlottenburg. The focus until then had been on the Parseval airships.

A short time later, LFG took over the "inheritance" of the "Flugmaschinen Wright Gesellschaft", which had gone into liquidation, and also began to manufacture aircraft, originally in Reinickendorf near Berlin on a site adjacent to the Luftschifferbataillon.

Initially it built Wright-type aircraft under license, then in 1912 it launched the first LFG-branded aircraft, which soon became known as "Roland" aircraft.

During the war, LFG was given the short name "Rol" by the Flugzeugmeisterei.

The LFG comprised various business facilities:
a) Plant in Charlottenburg (aircraft factory);
b) Bitterfeld plant (factory for airships and seaplanes built in 1908);
c) plant in Stralsund (factory for seaplanes);
d) an old hangar in Reinickendorf near Berlin (see above).

For the purpose of flying in the newly produced aircraft, a branch factory was established at the airfield in Berlin-Adlershof.

The shareholders of Luft-Fahrzeug-Gesellschaft m.b.H. included a number of banks, Bank für Handel und Industrie, Deutsche Bank and Nationalbank für Deutschland, as well as companies such as Elektrochemische Werke Friedrich Krupp, Ludwig Loewe and Hugo Stinnes.

Factory Facility and Airfield:

The old Wright works in Adlershof were expanded into a factory and by the end of 1915 had a working area of around 6,000 m². Further new buildings were in preparation when in September 1916 the entire factory in Adlershof, by now grown to about 20,000 m², burned down. Production was subsequently relocated to Charlottenburg to the large exhibition hall of the Verein deutscher Motorfahrzeug-Industrieller on Kaiserdamm. From a small parade ground a few hundred meters from the factory, the aircraft took off preferably only for takeoff to the airfield in Adlershof, 22 km away as the crow flies. There, the company had a few hangars to store the aircraft.

The buildings themselves were continually expanded and formed the aforementioned main branch in Charlottenburg. By the end of the First World War, the factory premises in Charlottenburg occupied an area of 70,000 m². All LFG plants together had an area of over 400,000 m² at their disposal.

Above: Map of Germany and Berlin-Charlettenburg.

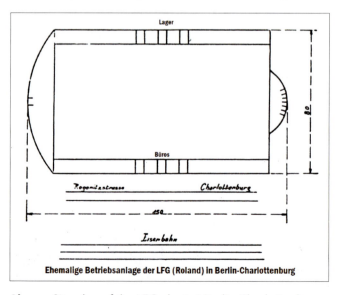

Above: Site plan of the LFG plant at Berlin-Charlettenburg.

Luft-Fahrzeug-Gesellschaft mbh (Roland)

Above: Aircraft factory in Charlottenburg (Berlin).

Above: LFG hangars in Adlershof.

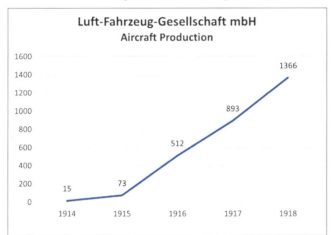

Above: Two Rol D.I and C.IIa in the production hall in Charlottenburg (former airship hangar).

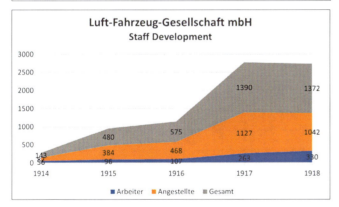

Staff and Designers:
The increase in the company's workforce, which developed in rapid succession, is shown in the adjacent chart.

Aircraft Development:
The LVG company was originally engaged in the construction of non-rigid airships, Parseval system, and in 1909 started building airplanes. In 1913, the Roland-Stahl-Pfeil biplane with a 100 hp Mercedes engine was built. The first aircraft of this biplane were made almost exclusively of metal, because the company wanted to undertake test flights in the colonies and wood was out of the question because of the climatic conditions. As early as the spring of 1913, a Roland steel biplane was taken to German Southwest Africa, which made numerous cross-country flights and excelled especially after the outbreak of the war until it was destroyed. During the war, it reverted to wooden construction due to the shortage of raw materials.

The company bought the Wright Company, which had gone into liquidation as a result of patent disputes, and in further development and using the existing facilities began building modern aircraft in 1912, for which it introduced the trademark "Roland" as a symbol of power and reliability. With these self-designed machines, it achieved not insignificant successes in 1913 and 1914 and received an incentive for further lively activity from the Nationalflugspende, which was int ended to secure Germany a leading position in aviation.

The first monoplanes and biplanes were made almost entirely of steel and were called "Roland-Stahl-Taube" and "Roland-Stahl-Doppeldecker" respectively.

At the outbreak of war in 1914, the LFG initially saw its main task in training young pilots.

Although Roland monoplanes and biplanes had already been delivered to the Army Administration, these types were

Aircraft Built by Luft-Fahrzeug-Gesellschaft mbH, Berlin-Charlottenburg (Rol)

Type	Engine [hp], Manufacturer	Crew	Armament	License Built By	Year	Notes
Landplanes						
Roland Steel-Arrow-Monoplane (A.I)	100, Argus As.I and others	2	–		1912	
Roland Steel-Arrow-Biplane	100, Merc. D.I and others	2	–		1913	
Alb B.II(Rol) [Rol B.I]	100, Benz Bz.II and others	2	–		1915	~280
Alb B.IIa(Rol)	120, Argus As As.II	2	–			550
Alb C.I(Rol) [Rol C.I]	160, Merc. D.III 150, Benz Bz.III	2	1 MG		1915/16	88
Rol C.II	160, Merc. D.III	2	1–2 MG	LiHo: 52	1916	~120
Rol C.IIa „Walfisch"	160, Merc. D.III	2	2 MG		1916	~85
Rol C.III	200, Benz Bz.IV	2	1 MG		1916	1, burned down
Rol C.V	160, Merc. D.III	2	2 MG		1916	1 prototype
Rol C.VIII	260, Merc. D.IVa 245, Mayb. Mb.IV	2	2 MG		1916	1 prototype
Alb C.X(Rol)	260, Merc. D.IVa	2	2 MG		1917	100
Han CL.II(Rol)	180, Argus As.III	2	2 MG		1918	200, order of 500
Rol D.I	160, Merc. D.III	1	2 MG	Pfal: 20	1916	~100
Rol D.II „Haifisch"	160, Merc. D.III	1	2 MG	Pfal: 200	1917	32
Rol D.IIa „Haifisch"	180, Argus As.III	1	2 MG		1917	140
Rol D.III	160, Merc. D.III 180, Argus As.III	1	2 MG	Pfal: 100	1917	100
Rol D.IV (Triplane Dr.I)	160, Merc. D.III	1	2 MG		1917	1 (2) prototype
Rol D.V	160, Merc. D.III 180, Argus As.III	1	2 MG		1917	3 prototypes
Rol D.VIa	170, Merc. D.IIIa	1	2 MG		1917	150 (+ 3)
Rol D.VIb	185, Benz Bz.IIIa	1	2 MG		1918	200
Rol D.VII	195, Benz Bz.IIIbo	1	2 MG		1918	prototype
Rol D.VIII	195, Benz Bz.IIIbm	1	2 MG		1918	prototype
Rol D.IX	160, S&H Sh.III 160, Goebel Goe.III	1	2 MG		1918	3 prototypes
Rol D.XIII	195, Körting Kg.III	1	2 MG		1918	Destroyed by fire
Rol D.XIV	160, Goebel Goe.III	1	2 MG		1918	1 prototype
Rol D.XV	170, Merc. D.IIIa 185, BMW BMW.IIIa	1	2 MG		1918	4 prototypes
Rol D.XVI (Rol E.I)	160, S&H Sh.III 160, Goebel Goe.III	1	2 MG		1918	2 prototypes
Rol D.XVII	185, BMW Bmw.IIIa	1	2 MG		1918	1 prototype
Rol G.I	245, Maybach Mb.IV	2	1 MG		1917	1 (burned, 2 propellers

Luft-Fahrzeug-Gesellschaft mbh (Roland)

Aircraft Built by Luft-Fahrzeug-Gesellschaft mbH (Rol)						
Type	Engine [hp], Manufacturer	Crew	Armament	License Built By	Year	Notes
Seaplanes [Plants in Bitterfeld (B) and Stralsund (S)]						
LFG Roland W1 (Alb C.Ia on floats)	150, Benz Bz.III	2	–		1915	1 prototype (B)
LFG Roland W (WD) (D.I on floats)	160, Merc D.III	1	2 MG		1916	1 prototype (B)
LFG Rol W16	160, Merc D.III	2	3 MG		1918	1 prototype (B)
LFG Rol V19	110, Oberursel U.II	1	–		1918	1 prototype (S)
Sab SF5	160, Merc D.III	2	–		1916	10 (S)
FF33(Rol)	160, Merc D.III	2	1 MG		1917	22 (S)
FF49C(Rol)	200, Benz Bz.IV	2	2 MG		1918	35 (S)

Above: Roland-Steel-Arrow-Biplane, 1913. here stationed in German Southwest Africa.

Above: Roland-Steel-Arrow-Monoplane (Taube) powered by Mercedes D.I engine, 1913.

Above: This Rol D.IIa became the 500th aircraft to be delivered.

Above: Fuselage structure of the Roland C.II "Whale", 1915/16.

not yet sufficiently war-tested to send them to the front. Therefore, on the advice of the Army Administration, LFG decided to build the proven Albatros aircraft under license. License construction then began rapidly at the beginning of 1915 under the type designation Alb B.II(Rol), equipped with 120 hp Mercedes engines. In further development, a reconnaissance aircraft without armament was built, and later a more powerful two-seater biplane Alb C.I(Rol), also an Albatros design, but with a rotating machine gun.

As a result of extensive tests conducted by the Göttingen Experimental Institute, the company produced an aircraft under the designation "Roland-Walfisch", which was characterized in particular by increased speed (by 30 km/h), improved field of fire and climbing ability. The Rol C.II

Above: Rol C.II muffler, ear radiator, propeller spinner.

Above: This Rol D.VI was the 1000 LFG-aircraft.

Above: The 2000th Roland aircraft was built in Charlottenburg: Roland D.VIb.

"Whale" aircraft achieved an epoch-making reduction in propulsive drag. This Roland C.II biplane (Walfisch) began its test flights in October 1915. However, the first plane already had an accident, because forced to land quickly by engine failure, it rolled against a bollard and broke the wing.

In 1916, the Roland biplane C.III with a 200 hp Benz engine was built, which was unfortunately destroyed by fire in the factory while still in the experimental stage. After this fire and the relocation to the newly leased factory premises on Kaiserdamm in Charlottenburg, a new design was created, a single-seater fighter called the "Haifisch" (Shark) under the type designation Rol D.II with a 160 hp Mercedes engine.

In February 1917, the company delivered its 500th aircraft and in October 1917, its 1,000th aircraft as the Roland biplane D.VI. By the end of the war, the production rate had risen to 10 aircraft per day.

Above all, efforts were made to reduce the harmful frontal drag of the aircraft to a minimum. This problem could not be solved with the practical experience of the aircraft builders alone. It was now necessary to apply the findings of science, in particular those of the Göttingen Aviation Research Institute. The low-drag drop shapes developed here were to form the basis of future aircraft fuselage shapes. Thus, the Rol C.IIa "Whale" developed by LFG was the first to emerge, which allowed the top speed to shoot up by 30 km/h.

At that time (1916), the emerging shortage of metals, especially steel, was already foreseeable. Skilled metal workers were tied up in other industries, such as the munitions industry. In contrast, carpenters and wood were freely available, which led to the "Whale" being made predominantly of wood. Even the stems, the basic framework of the hull, the frames, for which steel had also been used exclusively up to then, were now made of wood. This required a completely new strength determination for the entire aircraft. As a result, the Rol C.IIa was structurally better and lighter. External features, such as the propeller hub and ear cooler and the teardrop-shaped muffler, promised aerodynamic progress.

Four weeks after the catastrophic fire at the Adlershof plant (September 6, 1916), the move to the hangars on Kaiserdamm (Charlottenburg) was completed. On October 9, 1916, the first aircraft left the factory there. Several hundred "Walfische", now no longer built solely by LFG itself but under license from several major aircraft factories, reached the front. The German aircraft industry made the most imaginable efforts to catch up with LFG's lead. The shape of the Rol C.IIa became the model for most of all subsequent aircraft designs.

Development now proceeded at a rapid pace. The "Walisch" was followed by the Rol D.II "Haifisch" as a

Luft-Fahrzeug-Gesellschaft mbh (Roland)

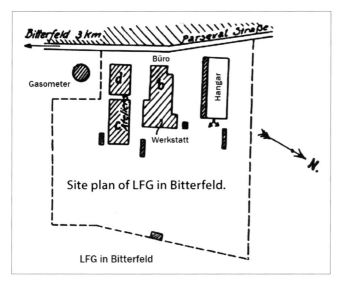

Site plan of LFG in Bitterfeld.

Above: LFG V19 single-seat low-wing monoplane from 1918 for use from submarines.
Left: LFG V19 single-seat low-wing monoplane from 1918 for use from submarines.

Above: Components of the V19, packed in individual containers for submarine carriage.

Above: It was envisaged that the V19, packed in individual parts, could be carried in the submarine. The aircraft was to be assembled on the deck of the submarine. Far left: Engine with fairing as a separate individual part.

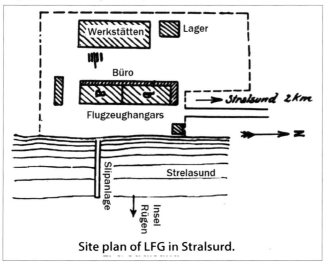

Site plan of LFG in Stralsurd.

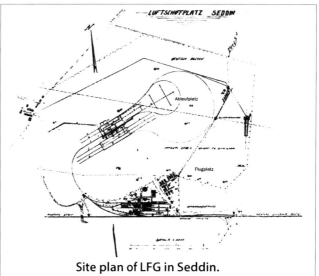

Site plan of LFG in Seddin.

single-seater fighter. In February 1917, the 500th aircraft was delivered, and in October 1917 the 1000th. An aircraft left the Kaiserdamm plant every two hours.

Since the outbreak of the war, LFG has built three unarmed two-seaters, seven armed two-seaters, twelve combat single-seaters - including a triplane -, and one large aircraft.

Plant Bitterfeld:

Airships, tethered balloons, motorized balloon cable winches and also seaplanes were manufactured in Bitterfeld. The immediate proximity to the chemical industry favored this location, especially for airship construction. From here, airships were not only supplied to the German army, but also exported to the allied powers of Austria and Turkey.

In 1916, the seaplane construction department of the LFG began operations in Bitterfeld, where the float version of the Albatros reconnaissance plane Alb C.I was initially built. This seaplane was released under the designation Lfg W. Final assembly and flying-in took place at Stralsund. After the end of the war, engineers Karl Theiß and Baatz designed 37 types, of which 17 were prototypes and three (V 13, V 60 and V 130) were built in small series.

The company was also engaged in the design and manufacture of large aircraft of the type Rol. G.I. Under construction was a 1,000 hp giant machine, but its components were destroyed by fire. Originally, it was planned to resume the construction of large aircraft, but this was not carried out due to the end of the war.

Plant Stralsund:

In Stralsund, up to 385 workers were employed in the production of seaplanes.

In front of the south hangar was a slipway for launching the aircraft.

Ten Sablatnig SF5s and 20 to 35 Friedrichshafen FF 49Cs were also manufactured under license at Stralsund. After the end of the war, engineers Karl Theiß and Baatz designed 37 types, of which 17 were prototypes and three (V 13, V 60 and V 130) were built in small series.

The LFG plant in Stralsund produced floats for "Friedrichshafen" seaplanes; but also 79 different seaplanes with boat hulls or on floats, and also small seaplanes designated as submarine aircraft (e.g., the LFG V19). These seaplanes were demountable and could be transported aboard the large submarines to which they served as reconnaissance aircraft. They were single-seat monoplanes with an 80 hp rotary engine, made entirely of duralumin.

The LFG Stralsund V19 Putbus was completed in September 1918 and conducted flight tests on behalf of the Imperial German Navy until the Armistice. Three production V 19s were ordered, but none were built by the time the Armistice was signed in November 1918. Interestingly, the Putbus was spared from demolition and scrapping under the terms of the Inter-Allied Control Commission and continued to fly until 1923, when it was eventually scrapped after failing to find a commercial role.

In 1919, the LFG factory in Stralsund produced two single-seat civilian seaplanes with boat hulls, each with a 100-horsepower engine. In 1920, aircraft production was discontinued, and the factory produced engine-powered wooden boats for fishing.

There were other facilities in Seddin, as well as a hangar in Reinickendorf (without seaplane production). The Seddin airship port was established during the First World War. It was located at the Seddin outworks near Jeseritz in Stolp County, Pomerania. It consisted of a landing field for airships, two airship hangars, a hydrogen production plant, warehouses and workshops.

During the war, Seddin Station was an important airship station for the German Navy for operations in the Baltic Sea and on the Russian border.

The factory facility in Reinickendorf, near Tegel, north of Berlin, near the tethered balloon center, consisted of a small wooden shed similar to the one in Bitterfeld. A small car for drag tests was found there.

Production supervisors (Bauaufsicht): #5

Construction Supervision No. 5, established at the company in September 1916, was located near the main factory in Charlottenburg. For the acceptance of the aircraft, part of the construction supervision was relocated to the Adlershof site.

The personnel of the construction supervision consisted of two officers, one engineer and one foreman each, and another 16 men.

Supervision of the Stralsund workshop was organized by the Imperial Navy.

War Material Found by the Inter-Allied Commission of Control (IACC) During Inspections After the End of the War:

No war material was found on the grounds of the LFG in Berlin-Charlottenburg during the visits of the Interallied Control Commission.

Situation after WWI:

In 1919, the Roland factory produced metal beds, agricultural machinery, electric lighting equipment and various furniture. About 500 workers remained at the plant for these different jobs.

In March 1920, the Roland factory again became the property of the Verein Deutsche Motor-Fahrzeuge, but was closed again in January 1921.

The former director of "Roland Maschinenbau" founded a new company: Roland Holz und Metallverarbeitung GmbH (with headquarters in Charlottenburg and a factory in Biesenthal/Mark).

Above: Roland Steel-Arrow Biplane in flight (1913)
Left: Roland-Stahl-Taube (1913)

Luft-Fahrzeug-Gesellschaft mbh (Roland) 245

Above: Roland C.II (1916)
Right: Roland C.IIa „Walfisch" (1916)

Above: Roland C.III (1916)

Above: Roland C.V (1916)

Above: Roland C.VIII (1917)

Above: Roland D.I (1916)

Above: Roland D.II "Haifisch" (1916)

Above: Roland D.IIa (Haifisch) (1917)

Development of Aircraft Factories

Above: Roland D.III (1917)

Above: Roland D.IV [Dr.I] (1917)

Above: Roland D.V (1917)

Above: Roland D.VI (1917)

Above: Roland D.VIa (prototype with modified fin and rudder). (1917)

Above: Roland D.VIb (prototype with dual I-struts). (1917)

Above: Roland D.VII (1918)

Above: Roland D.IX (S&H Sh.III) (1918)

Luft-Fahrzeug-Gesellschaft mbh (Roland)

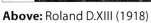

Above: Roland D.XIII (1918)

Above: Roland D.IX (Goebel Goe.III) (1918)

Above: Roland D.XIV (1918)

Above: Roland D.XV – Parallel struts, Merc. D.IIIa (1918)

Above: Roland D.XV – I-struts (1918)

Above: Roland D.XV – N-struts, 185 BMW.IIIa (1918)

Above: Roland D.XVI (Rol E.I) (160 hp Goebel Goe.III) (1918)

Above: Roland D.XVI (Rol E.I) (160 hp S&H Sh.III) (1918)

Above: Roland D.XVII, 185 Bmw.IIIa (1918)

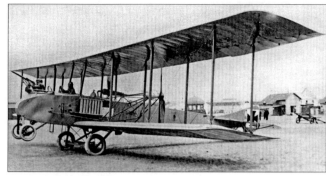
Above: Roland Rol G.I (1917)

Above: Roland W (Alb C.Ia on floats) (1915)

Above: Roland WD (Rol D.I on floats) (1916)

Above: Roland W16 – Seaplane (1918)

Above: Roland V19 (1918)

Above: FF49C(Rol) 1842, the first of 15 of this Roland-built batch.
Right: FF33S 3019. Roland built 22 of this type (MNs 789–790 & 6501–6520) under license.

Luft-Fahrzeug-Gesellschaft mbh (Roland)

Above: Albatros B.II(Rol) [Rol B.I] (1914/15)

Above: Albatros B.IIa(Rol) (1917)

Above: Albatros C.I(Rol)

Above: Albatros C.X(Rol) [Rol C.X]

Above: Halberstadt CL.IV(Rol)

Above: Hannover CL.II(Rol)

Above: Hannover CL.IIa(Rol)

Above: The Roland C.II was Roland's more successful design.

4.22 Flugzeugwerft Lübeck-Travemünde GmbH, Travemünde

Foundation:
In the course of the expansion of the Deutsche Flugzeugwerke (DFW) plant, they founded the Flugzeugwerft Lübeck-Travemünde in 1914 at the initiative of its director, Kommerzienrat Bernhard Meyer, together with the city of Lübeck and its financial support. The Flugzeugwerft Lübeck-Travemünde was entered in the commercial register on June 8, 1914.

A 5-year contract between the finance directorate of the city of Lübeck and the "Flugzeugwerft Lübeck-Travemünde GmbH," newly founded by the entrepreneur Bernhard Meyer of the Deutsche Flugzeugwerke Leipzig, initiated the start of aviation and the construction of an airport on the Priwall, a peninsula about three kilometers long at the mouth of the Trave River in eastern Schleswig-Holstein. This had been preceded by explorations to find a combined land and seaflight site on the Baltic Sea suitable for the German Reich. Priwall was given preference over Warnemünde for geographical reasons.

In August 1917, after the death of Kommerzienrat Meyer, the aircraft company was taken over by Fokker. In late summer 1918, however, Karl Caspar was in negotiations with Anthony Fokker regarding the takeover of Flugzeugwerft Lübeck-Travemünde GmbH. The company, located in Travemünde on the Priwall, had originally belonged to the Deutsche Flugzeug-Industrie GmbH group in Leipzig (DFW), along with Deutsche Flugzeugwerke GmbH in Leipzig-Lindenthal and National-Flugzeugwerke Allgemeine Flug GmbH in Berlin-Johannisthal. Within this group, the Lübeck-Travemünde Flugzeugwerk specialized in the production of seaplanes. It thus became a branch of Hanseatische Flugzeugwerke Caspar AG Hamburg (HFC). The sole shareholder was Dr. Karl Caspar (see also Section 4.16).

Factory Facility and Airfield:
In German aviation industry circles, the favorable location of the Priwall peninsula at the Pötenitzer Wieck was recognized early on: The inland sea-like bulge of the eastern shore at the mouth of the Trave River with its water surface of 3000

Above: Map of Germany and Travemünde.

Site Map of Flugzeugwerft Lübeck-Travemünde.

Below: Flugzeugwerft Lübeck-Travemünde (ca. 1917)

Above: Airfield and hangar of Flugzeugwerft Lübeck-Travemünde.

Above: DFW B.I aircraft visiting the Flugzeugwerft in Travemünde.

meters north-south extension and 1700 meters west-east extension, as well as with a depth of up to nine meters, offered ideal conditions for the developing maritime aviation industry. According to the plans of architect Max Bischoff, three wooden, unusually large assembly halls were built for the yard, as well as an administration building.

Fabrication:
On June 5, 1914, the first aircraft of the Travemünde Flying School took off from Priwall. These were a domestic aircraft, albeit a land plane, which were used for officer training purposes. As a result of the frequent accidents, the shipyard was initially well occupied with repair orders.

Kommerzienrat Meyer, who had been managing the "Deutsche Flugzeugwerke GmbH" (DFW) in Leipzig since October 30, 1911, manufactured already about 100 aircraft per year there (see also chapter 4.8). From June 8, 1914, just a few weeks before the outbreak of war, the aircraft plant in Lübeck-Travemünde was officially ready for operation. The following technical departments were set up: Fuselage construction, wing construction, and float construction.

In the early months, the untrained Lübeck workers were instructed by employees of DFW Leipzig-Lindenthal. Most of the work was repair work and the assembly of aircraft from the mother plant.

First assembled were the Rumpler-Etrich Taube built under license by DFW, then the DFW D.III and Brand B.II(DFW). The few employees at the Lübeck Flugzeugwerft produced just over 20 seaplanes for the Kaiserliche Marine by the end of the war, a number that was extremely small compared to the total.

Staff and Designers:
In 1914, the aircraft factory had just 18 employees, but as the First World War progressed, the number increased considerably - 189 employees by the end of the war.

Aeronautical Development:
The Deutsche Flugzeugwerke (DFW), which emerged from the Sächsische Flugzeugwerke in 1911, was first confronted with the problems of seaplanes in 1913 when it received an order from the Kaiserliche Marine to build a flying boat. The

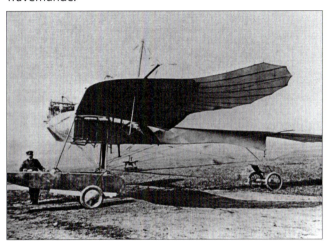

Above: First Navy aircraft: Taube on floats. The Fritzsche monoplane took off from Sonderburg, on the island of Alsen, with First Lieutenant Löw as pilot and Captain Lieutenant Busch as guest on its overseas flight to Kiel on June 18, 1911.

prototype, built according to the client's specifications, was a biplane of wooden construction with a 100 hp Mercedes engine. Nothing is known about successful flights, but the company looked forward to new orders with confidence. In 1916, orders from the Navy followed, prompting DFW to establish a new aircraft plant with access to the water. This took place in Travemünde near Lübeck.

In 1917, the first seaplane designed by Fritsch was introduced - the F.1. This aircraft was a DFW D.III equipped with floats. All three aircraft (Navy Nos. 282 - 284) were unique, because first the wings were staggered, on the next one they were unstaggered, and the latter was swept (similar to DFW's Mars biplane). Only the Navy No. 282 was delivered, as there were delays in the first F.1 due to various improvement requests from the Navy.

The F.2 (prototype: Marine No. 677), which had been ordered in 1916, even before delivery of the F.1, did not make its maiden flight until 1917. It was the first armed seaplane to be manufactured at Travemünde. The F.2 was

Development of Aircraft Factories

Aircraft Built by Luft-Fahrzeug-Gesellschaft mbH (RoI)

Type	Engine [hp], Manufacturer	Crew	Armament	License Built By	Year	Notes
F.1	160, Merc D.III	2	–		1917	1 (3)
F.2	220, Merc. D.IV	2	1 MG		1916/17	16 (17)
F.3	150, Benz Bz.III	1	–		1918	1
F.4	200, Benz Bz.IV	2	2 MG		1918	3 (30)

Above: Lübeck-Travemünde F.1 prototype.

Above: Prototype Lübeck-Travemünde F.2 with 4-bay wing.

Above: Prototype Lübeck-Travemünde F.3.

Above: Lübeck-Travemünde F.4.

also manufactured in various configurations. According to refernces, ten aircraft were delivered with the carrying deck raised for better pilot visibility from April 1917 to February 1918 (Marine Nos. 1147–1156). A second order for 9 aircraft was placed in October 1917 (Marine Nos. 1971–1979), with #1971–1973 being delivered as F.4s.

The two-seater biplane with a wingspan of about 19 meters, based on the fuselage of the DFW B-types, was equipped with amphibious floats. The designer was Heinrich Oelerich[1]. The powerplant was an eight-cylinder 220 hp (164 kW) Mercedes D.IV engine with transmission and a four-blade wooden propeller. The crew consisted of a pilot and an observer in tandem seating arrangement. The observer's seat was equipped with a 7.92 mm Parabellum machine gun mounted on a swivel turret.

Eleven F.2 aircraft were used by the Imperial Naval Forces of the German Empire. The aircraft were stationed at the Travemünde-Priwall seaplane base, which was also the location of the Imperial Navy's flying school.

The following F.3 was only built as a single unit. The only aircraft was not accepted by the SVK (Seeflugzeug-Versuchskommando) on March 23, 1918.

The F.4 was the flagship of the Lübeck-Travemünde aircraft plant. In addition to the three aircraft already mentioned, a further 30 aircraft of this type with Marine Nos. 7001-7030 were delivered by November 1918. It is doubtful whether all aircraft were actually completed.

One reason for stopping the production of the F.4 aircraft may have been that the new owner of the company, Karl Caspar, signed a license agreement with Flugzeugbau Friedrichshafen (FF) for the construction of the FF.49C seaplanes, which were well established within the German Navy. Some of these aircraft were not completed and sold as civilian aircraft until after the end of the war.

War Material Found by the Interallied Aeronautical Control Commission in Germany (IACC) During Inspections After the End of the War:

Four airplanes, two seaplanes, and a certain number of floats made of aluminium, fuselages of training airplanes, plans and various spare parts.

Lübeck-Travemünde

Situation of the Plant after the End of the War:
After the armistice in 1918, the Lübeck-Travemünde Flugzeugwerft continued the production of aircraft, namely several Travemünde F.4 machines as well as 10 licensed aircraft of the Friedrichshafen FF.49C type. The latter had been ordered by the naval administration as late as 20.9.1918 (Marine no 7031-7040) and were already converted to "civil" during the construction phase. Towards the end of 1918, the company had a staff of 189; by the end of 1919, the number had fallen to 91 workers.

On January 10, 1920, the Treaty of Versailles came into force, banning aircraft construction in Germany. But things were different in Lübeck-Travemünde. On April 8, 1920, the factory was renamed "Caspar-Werke GmbH". In the new name, the addition "Flugzeugwerke" was deliberately omitted, because aircraft construction was, after all, formally banned in Germany by the Treaty of Versailles. Officially, household appliances and furniture were manufactured.

A few weeks later, another change took place that was probably even more significant: As of April 8, 1920, it was stipulated in the articles of association that the headquarters of the Caspar group of companies was now Caspar-Werke mbH, Lübeck-Travemünde, and the parts of the former Hamburg HFC plant that remained with Caspar were downgraded to the status of a branch office in Hamburg.

The chief designer was Ernst Heinkel for a little more than a year. Aircraft were now also being built again, and were even delivered to the USA. After the change of name, around 30 types of aircraft were developed in Travemünde until its final closure in 1930, and 25 of them were built.

In 1923, 10 aircraft were still being built, but the low order situation led to economic ruin; even the orders from the Reich Association of the German Aviation Industry to test seaplanes could not save the factory. The Reichsmarine, in the person of Captain Paul Lohmann, gradually acquired the shares of Caspar-Werke AG until 1927 and brought all the parts into the Travemünde Naval Test Center.

Endnote:
1 Heinrich Bernhard Oelerich (* February 5, 1877, in Hamme; † March 23, 1953 in Freising) was a German aircraft designer and automobile racer. After receiving his German pilot's certificate No. 37 on October 21, 1910, at Berlin-Johannisthal on a Schultze-Herfort monoplane,[3] Oelerich made show flights in England, France, Portugal and South America. He then joined the Deutsche Flugzeug-Werke (DFW) in Lindenthal as chief pilot. Here he flew several records, such as a continuous flight over 2 h 41 min on July 5, 1912, with two passengers, a long-duration flight over 6 h 8 min on July 8, 1913, and a world altitude record over 8000 m on July 14, 1914. During World War I, Oelerich was a factory pilot, designer and technical director in the DFW, was involved with Hermann Dorner in the design of the DFW B-types and the DFW C.V, and participated in the work to develop the R-airplanes (Riesen-Flugzeuge). In 1918, he crashed during testing of the four-engine R.XV and from then on was severely handicapped.

Above: Prototype Lübeck-Travemünde F.1.

Above: Lübeck-Travemünde F.2 of first production batch.

Above: Lübeck-Travemünde F.2 of second production batch. The vertical tail resembled that of the F.4. The wings of all production F.2 aircraft had 3-bays.

Above: Prototype Lübeck-Travemünde F.4.

4.23 Lufttorpedo-Gesellschaft (LTG), Berlin-Johannisthal (Torp)

Foundation:
The Lufttorpedo-Gesellschaft Berlin, founded around 1916/17, worked mainly as a supplier for the aviation industry and generally produced wings, rudders, tension towers and tail units. The director of the Lufttorpedo-Gesellschaft, Max Schüler[1], contributed a patent for a sliding torpedo to the company when it was founded.

The military abbreviation was Torp.

Factory facility and airfield:
The aircraft hangar used by the LTG was located in the northern area of the Johannisthal airfield. It is not known where exactly the test flights of the seaplanes took place. It can be assumed that the Havel River was used for these activities.

Fabrication:
Only a seaplane type "SD1" (sometimes also called FD1 - See-D-Flugzeug) was developed. Of an announced land aircraft, it is not known whether it was built at all.

Staff and Designers:
The number of staff at the time of the reopening in 1917 was only 6. It is not known to what extent the number of staff increased during the war.

Aircraft Development:
From the beginning, a design department was also established, which developed two types of aircraft during the war: an unspecified biplane with wheeled undercarriage, built only in small numbers (if at all), and the FD1 fighter on floats. On February 8, 1917, the Navy Department placed an order with LTG for the construction of 6 of the latter, although the prototype was destroyed during the mandatory strength tests at SVK.

The LTG received the order for the three prototypes (Marine numbers 1299–1301) of a floatplane fighter on 8 February 1917. The first airframe for static testing was delivered three months later and was destroyed during the process. The next aircraft was delivered in July; flight testing showed that it was not very maneuverable and lacked longitudinal stability, so the third prototype was returned to the factory on 7 September.

LTG therefore revised the design and extended the keel and fin forward by about twice their length to improve stability in particular, whereupon the Navy ordered three more SD1s, which were assigned numbers 1518 to 1520. MN 1518 was delivered to the SVK on March 4, 1918, and were accepted as early as the 8th of that month. The Imperial Navy ended up taking delivery of five of the aircraft, but did not use them. One was given a wheel landing gear on a trial basis. The sixth FD1 was used for static testing and was destroyed in the process.

The LTG SD 1 was a single-seat sea-fighter aircraft which had as a special feature an upper wing directly connected to the fuselage. According to the Kaiserliche Marine's aircraft group classification, it belonged to the type of single-seat

Above: Map of Germany and LTG.

floatplane with fixed MGs (ED).

The two single-stage floats were 5.00 m long, 0.60 m wide and had a capacity of 948 l.

Production Supervisors (Bauaufsicht): #41
Production Supervision No. 41 for Berlin subcontractors, headed by Lieutenant Schmidt and Dipl.-Ing. Runne, was also responsible for LTG.

War Material Found by the Inter-Allied Commission of Control (IACC) During Inspections After the End of the War:
All five surviving aircraft were in storage at Hage (Aurich County) where the Allies found them in December.

Situation after WWI:
The LTG was liquidated at the end of the war.

Endnote

1 Max Schüler had already set up a small "Aeroplan factory" in Johannisthal in 1909, which is considered the first company of its kind in Johannisthal. For financial reasons, the company was soon closed and did not reopen until wartime.

Lufttorpedo-Gesellschaft (LTG)

Type	Engine [hp], Manufacturer	Crew	Armament	License Built By	Year	Notes
SD1 (FD1)	150, Benz Bz.III	1	1-2 MG		1917	1+5; Marine numbers 1299-1301 and 1518-1520

Above & Left: Torp FD1 (Marine numbers 1299 - 1301).

Above & Right: Torp SD1 Marine number 1518 with new fin design to improve stability and maneuverability.

Right: Undocumented single-seater photographed in the LTG factory. This may have been a landplane development of the SD1 or a type developed in parallel. While the size of the vertical tail surfaces did not need to be as large as the floatplanes, it still looks too small for good stability.

4.24 Luftverkehrsgesellschaft mbH (LVG), Johannisthal & Köslin

Above: Advertisement 1915.

Above: Map of Germany and Berlin-Johannisthal.

Foundation:

Luft-Verkehrs Gesellschaft mbH (LVG) was founded by Arthur Müller even before the war (1910) with the aim of carrying out advertising trips with Parseval airships Parseval P.L. 6 and P.L. 9 "Charlotte".

When the company did not deliver the hoped for results after one year due to technical difficulties, the company started building Farman type airplanes under license.

But the time of dependence on others was not to last long. At the end of 1911, the German army administration became interested in aircraft. They began to realize that an important weapon was in the making, even if they could not yet remotely foresee the extraordinary importance that the aviation industry would have in the not too distant future. Now the LVG saw a wide sphere of activity ahead of it and was determined to exploit the possibilities. The designer Franz Schneider was appointed, who had worked for a long time in the Nieuport factory in Reims and enjoyed the reputation of being a particularly reliable and skilled aircraft builder.

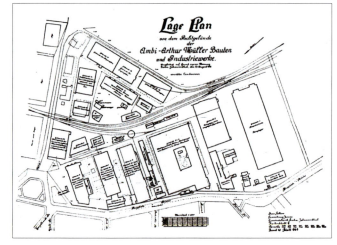

Above: Site plan of Berlin-Johannisthal LVG in 1920.

During the war, the Luftverkehrsgesellschaft mbH had been given the abbreviated designation "Lvg" by the

Luftverkehrsgesellschaft mbH (LVG)

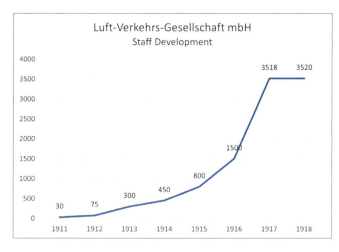

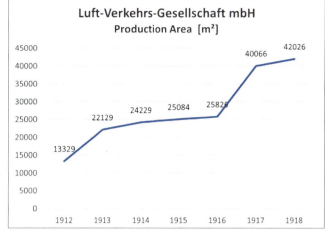

Flugzeugmeisterei. It owned:
a) A factory in Johannisthal along with its own screw factory in Berlin.
b) A factory in Köslin (Pomerania).

4.24a Luftverkehrsgesellschaft mbH, Berlin-Johannisthal, (LVG)

Factory Facility and Airfield:
The company was located at Johannisthal airfield and owned a factory site of approximately 53,000 m². The built-up operating area developed as follows (excluding outbuildings):

Fabrication:
Since its foundation, the LVG company was engaged in the design and manufacture of aircraft.

The following were produced: Farman type flying machines, LVG B.I, B II, C II, IV, V, VI, as well as some D and G aircraft.

The production figures developed over the years as follows:

1911	1 aeroplane
1912	6 aeroplane
1913	60 aeroplane
1914	50 aeroplanes per month
1915	85 aeroplanes per month
1916	60 aeroplanes per month
1917 (beginning)	70 aeroplanes per month
1917 (end)	137 aeroplanes per month
1918	156 aeroplanes

During the war, 5,640 aircraft were produced (including 1,100 C.VI aircraft from February to November 1918). Only Albatros Flugzeugwerke built more (6242 units). The highest monthly production rate achieved was 175 aircraft.

After the armistice and in 1919, 12 more Lvg aircraft were produced.

So, in total, at least 5,600 aircraft.

In addition, they delivered 3,923 boxes for machine gun cartridge strips.

The tables on the following pages show the main types of aircraft they designed or built under license.

Above: Advertisement 1915.

Staff and designers:
The number of employees has increased steadily since the year of its foundation. The growth can be seen in the following list:

The plant management was in the hands of a senior engineer, two plant managers and several plant engineers.

Aircraft Built by Luftverkehrsgesellschaft mbH (LVG)

Type	Engine [hp], Manufacturer	Crew	Armament	License Built By	Year	Notes
Landplanes						
LVG B.I (D 4)	100-120, Merc., Argus, Benz	2	–	Eul: 100 Ot: 20	1914	> 100
LVG B.II	110-120, Benz, Merc.	2	–	Schül: B.IIa: 200	1914/15	> 200
LVG B.III*	120, Merc. D.II	2	–	Eul: 100 Ssw: 100 Schül: 300 Hansa: 100 ?	1918	300
LVG C.I	150, Benz Bz.III	2	1 MG		1915	250
LVG C.II	150, Benz Bz.III 160, Merc. D.III 160, Mayb. Mb.III	2		Ago: 300	1916	> 700 3 on floats as D.IXw or C.IIW
LVG C.III	160, Merc. D.III	2	1 MG		1916	12
Alb C.III(LVG)	160, Merc. D.III	2	1 MG		1917	300
LVG C.IV*	220, Merc. D.IV	2	2 MG		1915/16	125
LVG C.V	200, Benz Bz.IV	2	2 MG		1917	1,250
DFW C.V(LVG)I	200, Benz Bz.IV	2	2 MG		1916–17	400
LVG C.VI	220, Benz Bz.IVa	2	2 MG		Jan 1918	1,000
LVG C.VIII	240, Benz Bz.IVaü	2	2 MG		End of 1918	1 prototype
LVG C.XI	160, Merc. D.III	2		Schül: 75		See Chapter 4.33c
Alb D.II(LVG) [LVG D.I]	160, Merc. D.III	1	2 MG		Ende 1916	75
LVG D.II (D 12)	160, Merc. D.III	1	–		1916	1 prototype
LVG D.III	185, Nag C.III	1	2 MG		1917	1 prototype
LVG D.IV	195, Benz Bz.IIIb	1	–		1917	2 prototypes
LVG D.V	195, Benz Bz.IIIb	1	–		1918	1 prototype
LVG D.VI	195, Benz Bz.IIIb	1	–		1918	1 prototype
LVG G.I	2x150, Benz Bz.III	3	1 MG		1915	1 prototype
LVG G.III	2x245, Mayb. Mb.IV	3	2 MG		1918	1 prototype
Go G.IV(LVG)	2x260, Merc. D.IVa 2x245, Mayb Mb.IVa 2x230, Hiero HIV	3-4	4 MG		1917	150 (+40 for Austria)
Go G.VII(LVG)	2x260, Merc. D.IVa	3	1 MG		1918	30
Go GL.IX(LVG)	2x240, Mayb. Mb.IVa	3			1918	70

Franz Schneider[1] - Chef designer. After Édouard de Nieuport's accidental death in September 1911, Schneider began working as technical director at Luft-Verkehrs-Gesellschaft A.G. in Johannisthal, where, in addition to three different monoplanes, he designed the LVG B and LVG C biplanes, which were used in large numbers by the German air force. In 1914, LVG sold two of these aircraft to Japan, and Switzerland also ordered six machines for the military; however, the outbreak of war prevented delivery.
In 1913, Schneider patented a firing device for firearms on aircraft to the German Imperial Patent Office (D.R.P. No. 276396). This patent used a locking mechanism that

Luftverkehrsgesellschaft mbH (LVG)

Aircraft Built by Luftverkehrsgesellschaft mbH (LVG)						
Type	Engine [hp], Manufacturer	Crew	Armament	License Built By	Year	Notes
Seaplanes						
LVG D.IXW (C.IIW)	160, Merc. D.III	2	1 MG		1915	3
Sab SF2	160, Merc. D.III	2	–		1916	10
Sab SF5	150, Benz Bz.III	2	–		1917	40
W.I [LVG C.IV on floats]	200, Benz Bz.IV	2			1915/16	1 prototype
W.II [LVG B.III on floats]	120, Merc. D.II	2			1918	1 prototype

* The LVG C.IV and LVG B.III, respectively, became seaplanes, types W I and W II, and were converted into civil transport seaplanes after the end of the war. A third seaplane existed under the designation D.9W.

Above: The D 3 airplane was a trainer built in 1912 with a 100 hp Argus or NAG engine. The nose was shrouded. The Prussian Army Administration ordered 18 aircraft of this type. A NAG engine was installed here.

Above: LVG E 2 monoplane, Schneider system from 1912. This type was tested with different 80–100 hp engines: Gnome, Schwade, or Oerlikon.

locked the gun's trigger via a linkage coupled to the engine's crankshaft when a propeller blade was in front of the muzzle. Schneider had this mechanism installed in the LVG E.I two-seat monoplane in 1915. The observer's seat was additionally equipped with a second, movable MG on a special rotating ring mount. Schneider also held a patent for this bogie. The LVG E.I was lost during transport to the front for unknown reasons, so no practical experience is available.

The locking device was further developed by engineer Heinrich Lübbe for Fokker Flugzeugwerke, using other patents, into a break gear ready for series production, which triggered a long legal dispute between Schneider and Fokker. As a result, Fokker patent No. 665528 of December 7, 1915, was cancelled by the Reichsgericht in Leipzig after the end of the war, and Schneider was awarded compensation in 1919. The principle of Schneider's rotating ring mount soon became standard in German military aircraft of the First World War.

It is also worth mentioning at this point that after the end of the war **August Euler** also received almost one million Reichsmark in license fees for his patent for the installation of a permanently installed machine gun in an aircraft, which had already been granted in 1912.

Ernst Heinkel[2] had been with LVG for about 15 months when he was approached by the famous flier Hellmuth Hirth (at the time Works Pilot for the Albatroswerke), and offered the position of Chief Designer for the firm. By this time Heinkel was anxious to use his own talents and not rely on the work of others; after he was promised a free hand in this respect by the Albatros partners Drs. Wiener and Huth, he accepted the job.

Heinkel's designs for Albatros followed one after the other with great rapidity. His first three aircraft were monoplanes, the second and third of the trio being of amphibian type. In the capable hands of Hirth contests were won and records broken almost every time the machines left the ground Heinkel was by now regarded as a first-class designer.

In April 1913, the first successful Albatros military aircraft was tested and accepted as the B.I. This machine was available with either 1, 2 or 3 bay wings as required and was the first of a long line of "B"-Class Albatros biplanes which were used as observation and training craft by the German Air Force during the four years of World War I.

In early 1914 **Igo Etrich** began seeking Senior Staff for the newly formed Brandenburg factory. Heinkel was approached and offered the position of Chief Designer at a salary he could hardly refuse. After some deliberation he accepted the position and, as mentioned earlier, commenced

Development of Aircraft Factories

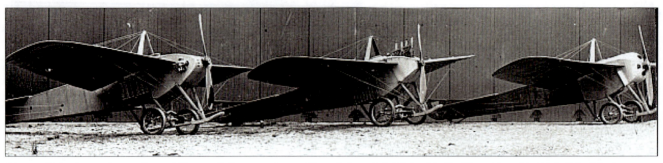

Above: This photo shows three different LVG monoplanes: (from left to right) E 2 with Oerlikon, E 1 with NAG, E 2 with Gnome.

his duties in March of that year at the Briest office.

Aircraft Development:

Since 1910, the company was engaged in the exploitation of guided balloons designed at that time for passenger and light advertising trips. It bought some Parseval airships and made the first trips with them in 1910. The company was attracted by the growing interest in airplane construction and turned its attention to this field as well. As early as 1911, the company presented two self-built biplanes based on the Farman system and one original Farman biplane for the B.Z. flight[3]. When the factory pilot and flight instructor Benno König[4] won the first prize on an Albatros-Farman biplane built by LVG on the very last day, the company became known to the general public for the first time. The construction of further airplanes, improved by own constructions, but always after the Farman system, was taken up and civil pilots, whom the company had accepted as pupils, were trained on it in flying.

In 1912, the first monoplane was built, based on the Nieuport type. Although several were built, this aircraft was not accepted by the army administration. After the flop of the monoplane, the company turned to the construction of a biplane based on the Farman system and produced a design on the basis of which the Army Administration signed an order for 18 units. Supported by these successes, another biplane was built, namely a fuselage biplane B I (internally D IV), which made its first test flights in May 1913 with very good results. This type gave rise to the so-called Prinz Heinrich machine, which performed brilliantly for its time at the Prinz Heinrich flight and successfully took part in the competition, winning six prizes.

The monoplane type, equipped with an in-line engine, was taken back into the program on an experimental basis, but was soon put aside permanently, as the results achieved were unsatisfactory. With seven airplanes of two types, the company successfully participated in the Ostmarken-Flug, as each participating airplane received a prize.

In further development of the above-mentioned biplane, the aircraft with the internal designation D IV was created at the outbreak of war in August 1914, which was initially considered one of the best aircraft types at the front. Subsequently, and in particular to meet the wishes of the army administration and the front-line units, the type C.I, a combat machine with a movable machine gun for the rear-mounted observer's seat, was created through redesigns and new constructions. Returning to the monoplane, this machine was again rebuilt as such, but after Lieutenant Wentsch's fatal crash it was finally discarded.

In February 1915, at the request of the Reichs-Marine-Amt, the company constructed a W.I experimental aircraft, a Lvg C.I with floats, for use as a torpedo plane. After completion in June of that year, the first flights were made and shortly thereafter the aircraft was acquired by the Navy as a training aircraft. Progressive improvements in the Lvg C.I aircraft resulted in the company receiving large series orders on the resulting new type, Lvg C.II, and in parallel also decided to build two large fighter aircraft, one with a fuselage and two tenders which served as engine foundations, and one with two fuselages in which the engines were installed, and a dinghy in the center to carry the occupants and armament. Since these aircraft did not meet the requirements set, construction of the large aircraft was abandoned. The same happened to a biplane with a wooden fuselage, which was put aside while still under construction because its fuselage was far too heavy. Further experiments in design and production led to the type Lvg C.IV (internally D XI), equipped with 220 hp 8-cylinder in-line engine Mercedes D.IV.

This type, which after various modifications had become a success, was followed in mid-1916 by an order from the Army administration to build 125 aircraft, which were mass-produced. The single-seat LVG aircraft, which was under construction in parallel, showed particularly good results in terms of top speed, which was still 206 km/h. Despite great expectations of this aircraft, they did not lead to the goal due to an unexpected wing failure.

After these unsuccessful attempts, the company initially refrained from further in-house developments. LVG made successful efforts to obtain licenses from the Army Administration to build Albatros Alb D.II and DFW C.V aircraft.

A machine with a 200 hp Benz Bz.IV engine designed during these license constructions, internally perceived as a Type D XIII, also led to no success. It was not until the fall of 1916, after Sabersky-Müssigbrodt had joined the company as designer and first technical director, that a new biplane LVG C.V (internally D XV) was built. This type was successfully accepted and went into serial production.

Luftverkehrsgesellschaft mbH (LVG)

In accordance with the wishes of the Inspektion der Fliegertruppen, LVG decided in September 1916 to resume construction of large fighter aircraft. This led to the start of series production of Gotha G.IV large fighters in early 1917. Further efforts to build its own large fighter did not bring any significant success.

In 1917, a fighter monoplane, aircraft type D.III, was started as a new design and sent to the front for test flights. Due to the fact that the N.A.G. C.III engine was too heavy, this type was abandoned as hopeless for better results. Subsequently, another single-seater was constructed for trials with a Benz high-speed V-shape engine.

A new training aircraft Lvg B.III was successful in 1918 and was built in larger series in the company's own factory and also under license from other companies. At the suggestion of the Flugzeugmeisterei, the company decided to build a C-airplane LVG C.VI with a 200 hp Benz engine Bz.IIIa, which started its first test flights in mid-1918. In addition to this type, another C-airplane C.VIII with over-compressed Benz engine Bz.IIIaü was constructed on the basis of the B.III training machine. However, this development was completed too late to be used.

Production supervisors (Bauaufsicht): #2

The construction supervision was established at the company on September 18, 1916. The leader of the construction supervision was an officer, to whom another officer and an engineer were assigned.

In addition to the above-mentioned, the personnel of the construction supervision consisted of another 27 persons, below which 10 were post-flight pilots. The high number of pilots is due to the large number of units.

War Material Found by the Inter-Allied Commission of Control (IACC) During Inspections After the End of the War:

129 aircraft and 74 engines along with a large quantity of spare parts.

Situation after WWI:

LVG dissolved itself in 1919. A limited partnership under the same name was then founded, the "Luft Verkehrs-Gesellschaft Arthur Müller Kommandit-Ges.". By the end of 1920, this company owned only a few offices in the former Lvg factory in Johannisthal and was engaged in technical research in the field of aviation.

By far the largest part of the LVG site was rented by the "Ambi" company in 1919 to produce a new type of mower and harvester. At the beginning of 1920, about 40 "Ambi" machines a day were coming off the production line.

Above: The E 4 was a racing monoplane with 80–100 hp Gnome, but also Oberursel rotary engine.

Right: Gotha GL.VII(LVG) (1918)

Left: LVG C.VIII in flight. The C.VIII was the ultimate two-seater design by LVG. Developed from the successful C.VI, the C.VIII used an over-compressed engine for better performance at high altitude, a nose radiator, and ailerons on all wings for improved maneuverability.

Above: LVG B.I with 145 hp V8 Rapp engine Rp.II, installed for test purposes.

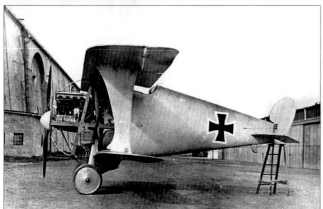

Above: "C hangar" of the LVG in Johannisthal.

Above: "G hangar" of the LVG in Johannisthal.

Left: Working on a Go G.IV(LVG).

Below Left & Below: The LVG D.10 with 120 hp Daimler D.II was an experimental aircraft with an extreme oval fuselage. Only one aircraft was built and tested in 1916.

Luftverkehrsgesellschaft mbH (LVG)

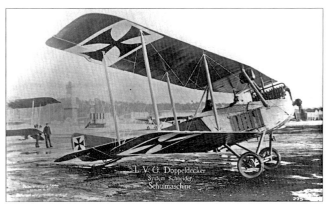

Above: LVG B.I (1914)

Left: LVG workshops and loading terminal.

Above: LVG B.II (1914)

Above: LVG C.I (1915)

Above: LVG C.II (1916)

Above: LVG C.III; observor in front cockpit with gun turret and flexible gun. (1916)

Above: LVG C.IV (1915/16)

Above: LVG C.V (1917)

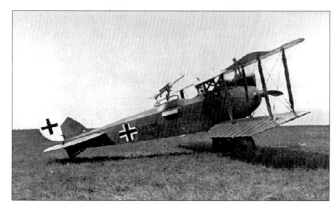

Above: LVG C.VI (1918)

Above: LVG C.VIII (1918)

Above: Albatros D.II(LVG) [LVG D.I] (1916)

Above: LVG D.II (1916)

Above: LVG D.III (1917)

Above: LVG D.IV (1917)

Above: LVG D.V (1918)

Above: LVG D.VI (1918)

Luftverkehrsgesellschaft mbH (LVG)

Above: LVG G.I (1915)
Right: LVG G.III (1918)

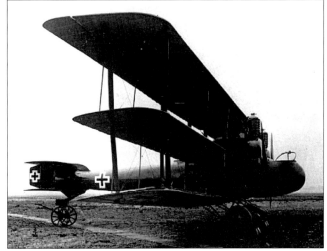

Above: Gotha G.IV(LVG) (1917)

Above: Gotha GL.IX(LVG) (1918)

Above: LVG D.IXw (C.II on floats) (1915)

Above: W.I [LVG C.IV on floats] (1918)

LVG-built Sablatnig SF2 (1916)

LVG-built Sablatnig SF5 (1917)

4.24b Luft-Verkehrs-Gesellschaft mbH, Köslin (now Koszalin, Poland) (LVG)

Foundation:
The company, a subsidiary of Luft-Verkehrs-Gesellschaft mbH Berlin (see also the previous section), was founded on September 1, 1915.

Factory Facility and Airfield:
The Köslin[5] aircraft factory comprised a larger factory site, which was directly connected to an airfield. Since its foundation, the site grew from just under 1,500 m² to over 11,000 m² at the end of the war.

Fabrication:
The company was initially engaged almost exclusively in the repair of aircraft of the Eastern Front, in particular of the types LVG C.II for the Inspection der Fliegertruppen and a number of sea planes of the type Sablatnig SF2.

In addition, the company produced about 40 new Sablatnig SF5 seaplanes until June 1917, when it received a larger order to produce 300 Alb C.III(Lvg) aircraft. At the same time, in parallel, about 100 repaired machines were returned to the front.

The production included over the years:
1915/16   60 new aircraft LVG C.II and 10 seaplanes.
1917   80 new aircraft Alb C.III(LVG), 40 seaplanes and 10 repair aircraft monthly.
1918   220 new aircraft Alb C.III(LVG) and 25 repair aircraft monthly.

Aircraft Development:
From the flight school established in Köslin in the spring of 1914, the branch factory was created due to the good local location of the airfield and favorable working conditions because of low labor costs.

After completion of the building erected for fabrication, repair work on Lvg B.I and B.II aircraft was started, until an order for the production of

Lvg B II front-line aircraft followed in January 1916. In April 1916, the company was awarded another order on 10 aircraft, Sablatnig type. As orders on front-line aircraft failed to materialize in September 1916, repairs were resumed in the workshops. In October 1916, LVG Köslin received a smaller order to supply seaplanes to the Reichs-Marine-Amt.

In May 1917, the parent company, having expanded enormously, placed an order with the Köslin branch for the delivery of 200 Alb C.III(LVG) aircraft, which was subsequently increased to 300 aircraft (with a capacity of 40 aircraft per month).

Staff and Designers:
The number of employees developed from 120 (early 1915) to about 680 in 1917/18.

Since the company exclusively carried out licensed construction and repair, no design department was established, nor were any design engineers employed.

War Material Found by the Inter-Allied Commission of Control (IACC) During Inspections After the End of the War:
123 aircraft and 63 Engines.

Above: Map of LVG plant in Köslin, now Koszalin, Poland.

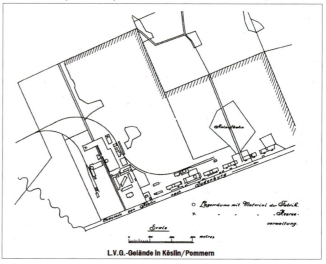

Above: Site plan of LVG plant in Köslin.

Situation after WWI:
In 1919, the factory "Lvg" in Köslin underwent numerous reconstructions in order to convert the production to peacetime production. By the end of 1920, the company employed about 300 workers who produced furniture and agricultural machinery.

Endnotes:
1 Franz Schneider (* September 27, 1871 in Constance; † May 24, 1941 in Tokyo) was a Swiss aviation pioneer and entrepreneur. Schneider had become a German citizen in the meantime, but retained his Swiss citizenship. Schneider left LVG at the end of 1916 after financial and legal disputes. From the liquidated Deutsche Eisenbahn-Speisewagen-Gesellschaft, Franz Schneider acquired the factory facilities in Seegefeld and on January 22, 1917, founded his own company, Franz Schneider Flugmaschinenwerke, with about 125 employees. Among the employees was the young Viktor Carganico, and the construction supervisor was Lieutenant Elchleb. Although

Luftverkehrsgesellschaft mbH (LVG)

Above: LVG branch in Köslin, top left halls of the flight school.

Above: Covering the surfaces and fuselages at LVG Köslin.

Schneider designed a single-seater fighter in 1918, the factory mainly repaired Albatros, DFW and LVG front-line aircraft. In 1919, Schneider and his family moved to Seegefeld. Aircraft were only allowed to be built in Germany under severe restrictions after the end of World War I, so the company tried to find new business. Around 1920, Schneider therefore changed the company name to Franz Schneider Maschinenwerke and, in addition to building and selling flying machines, now also offered railroad cars and machines of all kinds.

2 Ernst Heinrich Heinkel (* January 24, 1888 in Grunbach (Oberamt Schorndorf), Kingdom of Württemberg; † January 30, 1958 in Stuttgart) was a German engineer and aircraft designer and plant manager in several outside companies. After the war, E. Heinkel founded his own aircraft company. 3 The first "German Round Flight for the B.Z. Prize of the Skies 1911" was launched from Johannisthal airfield near Berlin on June 11, 1911. "B.Z." stands for "*Berliner Zeitung*".

4 Benno König (* June 16, 1885 in Untermenzing; † July 1, 1912 in Hamburg-Altona) was a German locksmith, chauffeur and aviation pioneer. On July 10, 1911, he won the first Deutschlandflug, a cross-country competition over 13 legs. About a year after his success in the Deutschlandflug, on June 30, 1912, his plane overturned after an engine failure over Langenfelde during an outlanding on a Chaussee. König was then injured and taken to Altona Hospital, where he died on the morning of July 1, 1912. His plane was a monoplane of his own design with a 65 hp Renault engine.

5 Meant is the present Koszalin, Poland, formerly German: Köslin, West Pomerania.

Above: LVG W.II [LVG B.III on floats] (1918). Built at LVG Johannistal, not Köslin.

4.25 Märkische Flugzeugwerft GmbH, Golm i. Mark (Mark)

Above: Advertising poster 1916.

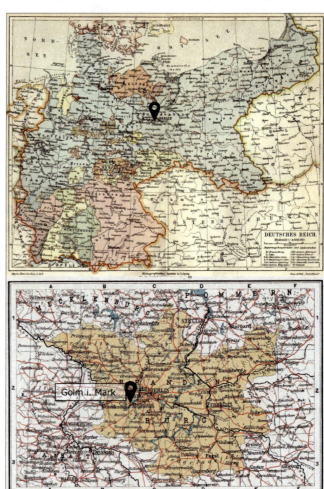

Above: Map of Germany and Märkische Flugzeugwerft.

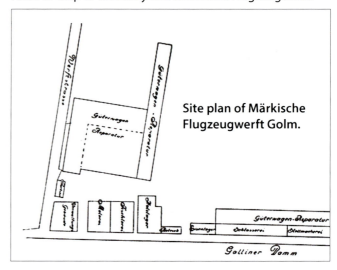

Site plan of Märkische Flugzeugwerft Golm.

Foundation:
The company was founded in Teltow on December 4, 1913 by Director Frank Eckelmann and Hans Coler. Initially, only test flights were carried out here. The construction of its own aircraft factory was not possible in Teltow. The Union Flugzeugwerke already located there prevented further expansion. In June 1914, the company's headquarters were moved to Golm i. d. Mark. Golm is now a district of Potsdam. In 1920, the company was renamed Märkische Industrie-Werke GmbH.

Factory facility and airfield:
The company owned a total area of nearly 54,000 m², of which 30,000 m² was devoted to the airfield and the rest to the production halls. The entire factory site was located west of Potsdam close to the water (Havel and Zernsee).

Fabrication:
While the company in Golm was initially concerned only with spare parts production, aircraft repair began as early as July 1914. The first repair machine was delivered in September of the same year. The number of repaired machines was increased to 28 by September 1916, with a workforce of 270. The company received its first order for the production of Rumpler type Ru C.I training aircraft in October 1916 and delivered the first machine in January of the following year. At the instigation of the Idflieg, a further

Märkische Flugzeugwerft (Mark)

Above: Märkische Flugzeugwerft in Golm. (1918)

Above: Several Rumpler Ru C.I(Mark) in production. The large numbers show the internal serial number.

Above: View at the Märkische Flugzeugwerft 1916.

Above: Aircraft hangars in Golm (1918).

expansion of the plant was undertaken so that the production of new aircraft could be increased in batches. By September 30, 1917, the 1916/17 fiscal year, the company had delivered 135 new aircraft. Further structural and mechanical changes made it possible to further increase production figures.
In January and February 1918, 80 new aircraft each were delivered. By the end of the war, the company was able to deliver up to 140 aircraft a month.

License production at Märkische Flugzeugwerft GmbH was limited exclusively to Ru C.I aircraft. Production was optimized to such an extent that up to 80 aircraft could be delivered per month. The defective aircraft were delivered from the front or the flying schools by rail and then a horse-drawn vehicle.

In addition, a military flying school was on site from 1918, but only 60 pilots were trained there.

Staff and Designers:
The company started in 1913 with 6 employees, in September 1916 there were already 270 and grew steadily to 1,700 men in 1918.

The director was Franz Eckelmann, the designer of the D I and D II single-seat fighter was Wilhelm Hillmann[1] Bruno Pöhlke[2] and Albert Schäfer were the pilots of the repaired airplanes of the aircraft works.

The company started in 1913 with 6 employees and grew steadily to 1,700 men in 1918.

Wilhelm Hillmann was active as design manager (see below). The test pilots were Albert Schäfer and Bruno Poelke.

Aircraft Development:
In addition to the licensed production of Ru C.I aircraft, the company was engaged in the finishing of two of its own single-seat fighters, originally designed by Wilhelm Hillmann of Schütte-Lanz.
Work was carried out from 1918 on:
D I - single-seat fighter with 195 hp Bz IIIbo. Interesting features were the drop tank under the pilot's seat and wing warping on the lower wings for lateral control. One prototype was built.

Aircraft Built by Märkische Flugzeugwerft (Mark)

Type	Engine [hp], Manufacturer	Crew	Arma-ment	License Built By	Year	Notes
Ru C.I(Mark)	150, Benz Bz.III	2	1 MG		1917	700
Mark D.I	200, Benz Bz.IIIbo (V8)	1	2 MG		1918	1 pre-series aircraft only
Mark D.II						

Above: Finishing aircraft repairs, Mercedes C.I, side radiator.

Above: Unhappy landing at the flight school of Märkische Flugzeugwerft.

D II - since Idflieg declined the wing warping, Hillmann developed an improved variant with ailerons, but still retained the wing warping! The development was not completed until the end of the First World War.

Background:
The Schütte-Lanz company, until 1918 mainly involved in the construction of airships, also developed a number of aircraft types at its Zeesen plant in Brandenburg, including the D.I and D.II projects.

The D.I was created in 1915 as a replica of the successful British Sopwith Tabloid as the first German fighter biplane, but was underestimated as a fighter and rejected, as the monoplane was considered more suitable as a fighter due to better visibility for the pilot. Its planned further development into the D.II with a more powerful 100 hp Mercedes in-line engine was therefore cancelled at Schütte-Lanz.

Production Supervisors (Bauaufsicht): #26
At the end of the war, the construction supervision at the Märkische Flugzeugwerke consisted of 14 persons, including 1 officer as chief and two non-commissioned officers as test pilots.

War Material Found by the Inter-Allied Commission of Control (IACC) During Inspections After the End of the War:
2 aircraft and 28 engines.

Situation after WWI:
After the war, Märkische Flugzeugwerft stopped being interested in aviation and changed her name. In 1920, it had been registered under the name: Märkische Industrie-Werke GmbH entered in the Commercial Register. From then on, it undertook repairs of railroad carriages and the manufacture of furniture.

Endnotes:
1 W. Hillmann acquired pilot license No. 559 in 1913, and had been working in the design department at Schütte-Lanz since the end of 1913. Before that, in the same year, he made the first ferry flight with an English Sopwith seaplane from England to Kiel. Later, he acted additionally for the Märkische Flugzeugwerft.

2 Bruno Poelke (* 1883 in East Prussia; † August 9, 1975 in Frankfurt am Main) was a German aviation pioneer from Frankfurt am Main. Poelke was one of the first to construct flying machines, along with the Wright brothers, several Frenchmen and Otto Lilienthal. Poelke reached an altitude of six meters above the point of ascent during a flight test from the gliding hill on July 9, 1909, the day before the official opening of the ILA. In 1910 he designed a biplane and in 1911 a low-wing monoplane, both of which he was one of the first to equip with a seven-cylinder rotary engine. In 1912 Bruno Poelke gave up his professional independence and flew the new types of various aircraft factories.

Above: Four aircraft Rumpler Ru C.I(Mark) shortly before being accepted by the local inspection authority Bauaufsicht #26.

Above: Three different aircraft types in the repair shop of Märkische Flugzeugwerft.

Märkische Flugzeugwerft (Mark)

Above & Below: Märkische D.I (1918)

Right: Rumpler C.I(Mark) (1917)

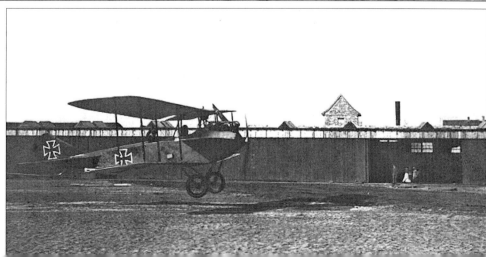

Made in United States
North Haven, CT
08 May 2024